Christian Gärtner

Organic Laser Diodes

Modelling and Simulation

Organic Laser Diodes

Modelling and Simulation

by Christian Gärtner

universitätsverlag karlsruhe

Dissertation, Universität Karlsruhe (TH)
Fakultät für Elektrotechnik und Informationstechnik, 2008

Impressum

Universitätsverlag Karlsruhe
c/o Universitätsbibliothek
Straße am Forum 2
D-76131 Karlsruhe
www.uvka.de

Universitätsverlag Karlsruhe 2009
Print on Demand

ISBN: 978-3-86644-345-7

Organic Laser Diodes: Modelling and Simulation

Zur Erlangung des akademischen Grades eines

DOKTOR-INGENIEURS

von der Fakultät für Elektrotechnik und
Informationstechnik der Universität Karlsruhe (TH)

genehmigte

DISSERTATION

von

Dipl.-Phys. Dipl.-Inform. Christian Gärtner

geb. in Karlsruhe

Tag der mündlichen Prüfung: 26.06.2008

Hauptreferent: Prof. Dr. rer. nat. Uli Lemmer
Korreferent: Prof. Dr.-Ing. Wolfgang Kowalsky

Karlsruhe 2008

Zusammenfassung

Im Rahmen der vorliegenden Arbeit werden die Grundlagen der Bauteilphysik von organischen Halbleiterlaserdioden unter hohen Anregungsdichten durch numerische Simulation untersucht. Organische Laserdioden versprechen eine Vielzahl einzigartiger Merkmale wie die Durchstimmbarkeit über das gesamte sichtbare Spektrum. Die Herstellungstechnologie ist prinzipiell sehr kostengünstig und organische Laser können außerdem großflächig auf flexiblen Substraten hergestellt werden. Diese Eigenschaften eröffnen organischen Laserdioden eine Vielzahl von Anwendungen in der Bioanalytik, im Digitaldruck, in der Fluoreszenz- Spektroskopie sowie für Sicherheitsanwendungen. Bis heute existieren jedoch nur organische Laser, welche mit Hilfe einer zweiten Laserquelle optisch angeregt werden. Optisch gepumpte Systeme haben jedoch aufgrund des zusätzlich benötigten Pumplasers größere Abmessungen und weisen wesentlich höhere Stückkosten auf, weshalb sie für viele Anwendungen nicht interessant sind.

Elektrisch angeregte Lasertätigkeit konnte bisher in organischen Materialien noch nicht gezeigt werden. In den vergangenen Jahren wurden einige Herausforderungen auf dem Weg hin zur Realisierung einer organischen Laserdiode erkannt. Die elektrischen Kontakte führen zu einer immensen Absorption innerhalb des Wellenleiters und somit zu einer erheblichen Dämpfung der Lasermode. Dieses Hindernis kann durch den Einsatz dicker organischer Transportschichten, welche als Wellenleitermantel fungieren, umgangen werden. Alternativ ist es auch möglich, schwach absorbierende Materialien als Elektroden (wie beispielsweise transparente leitfähige Metalloxide) einzusetzen. Durch eine geeignete Auslegung des Wellenleiters ist es möglich, die Propagation von Moden höherer Ordnung zu ermöglichen, wobei die Kontakte in die Minima des Intensitätsmodenprofils gelegt werden. Dadurch kann die Dämpfung des Wellenleiters auf etwa 3 /cm gesenkt werden.

Es wird erwartet, dass Stromdichten von etwa $1~\text{kA/cm}^2$ erforderlich sind, um elektrisch angeregte Lasertätigkeit zu erzielen. Organische Halbleiter weisen jedoch sehr viel geringere Ladungsträgerbeweglichkeiten auf als ihre anorganischen Pendants. Während für Galliumarsenid die Ladungsträger Beweglichkeiten im Bereich von $10^4~\text{cm}^2/\text{Vs}$ aufweisen, ist die Beweglichkeit der Ladungsträger in organischen Halbleitern im Bereich von 10^{-6} bis $10^{-3}~\text{cm}^2/\text{Vs}$. Deshalb sind elektrische Felder von mehr als $10^6~\text{V/cm}$ und Ladungsträgerdichten von bis zu $10^{19}~\text{/cm}^3$ erforderlich, um Stromdichten von $1~\text{kA/cm}^2$ zu erreichen. Werden organische Halbleiter unter derart hohen Anregungsdichten betrieben, treten zahlreiche Verlustprozesse in den Vordergrund, die bei organischen Leuchtdioden nur eine untergeordnete Rolle spielen, da diese bei weitaus niedrigeren Anregungsdichten betrieben werden. Bisher war nicht bekannt, welchen Einfluss diese Verlustmechanismen auf die Laserschwelle einer organischen Laserdiode haben, und welche Optionen zur Senkung des Einflusses der Verlustmechanismen existieren.

Im Rahmen dieser Arbeit wird der Einfluss von zahlreichen Verlustprozessen auf die Laserschwelle untersucht. Hierfür wird eine numerische Simulationssoftware eingesetzt, welche die Injektion von Ladungsträgern von den Elektroden in ein organisches Multischichtsystem, den Ladungstransport in Drift-Diffusionsnährung, die Rekombination von Elektronen und Löchern zu Exzitonen sowie die Wellenleitung der Lasermode abbildet. In diese Software wurde die Beschreibung von Verlustprozessen in organischen Materialien unter Hochanregung eingearbeitet. Mit Hilfe dieses Modells wurde die Ladungsträgerdynamik in organischen Doppelheterostrukturen sowohl im Gleichstrombetrieb als auch im Impulsbetrieb charakterisiert. Besondere Beachtung fand die Charakterisierung des simultanen Einflusses unterschiedlicher Verlustmechanismen. Im Detail wurden bimolekulare Annihilationsprozesse, feldinduzierte Exzitonendissoziation sowie induzierte Absorptionsprozesse durch Polaronen und durch Exzitonen im ersten angeregten Triplettzustand untersucht. Mit Hilfe der gewonnen Erkenntnisse konnten Design-Regeln für organische Laserdioden erarbeitet werden, welche eine Reduzierung des Einflusses der aufgezählten Verlustmechanismen erlauben. Darüber hinaus wurden unterschiedliche Entwurfskonzepte für organische Laserdioden in der OLED-Geometrie hinsichtlich der Empfindlichkeit gegenüber Verlustprozessen evaluiert.

In Abwesenheit von induzierten Verlustprozessen wurde für eine optimierte Doppelheterostruktur eine Laserschwelle von 206 A/cm^2 ermittelt. Bei Stromdichten dieser Größenordnung treten in organischen Bauteilen jedoch zahlreiche Verlustprozesse in den Vordergrund. Unter Berücksichtigung von bimolekularen Annihilationsprozessen wird die Laserschwelle erst bei einer Stromdichte von 8,5 kA/cm^2 erreicht. Hierbei zeigen Bauteile mit dickeren Emissionsschichten und gleichen Beweglichkeiten für Elektronen und Löcher eine geringere Empfindlichkeit gegenüber bimolekularen Annihilationsprozessen. Allerdings gelingt es bisher technologisch nicht, die Dicke des Bereichs, in dem Exzitonen gebildet werden, auf über 20 nm anzuheben. Gelänge es, die Breite des Bereichs in organischen Laserdiodenstrukturen, in dem Exzitonen homogen generiert werden, von 20 nm auf 100 nm anzuheben, würde die Laserschwelle bereits bei einer Stromdichte von 2,3 kA/cm^2 erreicht werden.

Wird gleichzeitig zu bimolekularen Annihilationen auch noch der Einfluss der feldinduzierten Exzitonendissoziation berücksichtigt, wird für die gewählten Materialparameter Lasertätigkeit verhindert. Bei Stromdichten oberhalb von 10 A/cm^2 führt die verstärkte Dissoziation von Singulett-Exzitonen, welche in organischen Materialien diejenige Teilchenspezies ist, welche den optischen Gewinn generiert, zu einer Akkumulation von Polaronen. Bei einer Stromdichte von 10 kA/cm^2 ist die Polaronendichte um etwa eine Größenordnung erhöht gegenüber der Berechnung ohne Exzitonendissoziation. Die erhöhte Polaronendichte führt zu einem zehnfach stärkeren Einfluss von Singulett-Polaronen-Annihilation, wodurch die Dichte der Singulett-Exzitonen stark verringert wird. Während bei einer Stromdichte von 10 kA/cm^2 unter alleiniger Berücksichtigung bimolekularer Annihilationsprozesse noch eine Singulett-Exzitonendichte von 3×10^{17}/cm^3 erreicht wird, beträgt diese nur noch 5×10^{15}/cm^3, wenn zusätzlich noch der Einfluss von feldinduzierter Exzitonendissoziation einbezogen wird. Gast-Wirt-Systeme, bei welchen ein Emitter mit kleiner Bandlücke in ein Matrixmaterial mit großer Bandlücke dotiert wird, weisen eine geringere Empfindlichkeit gegenüber Exzitonendissoziation auf. Bei Verwendung des Gast-Wirt-Systems Alq$_3$:DCJTB als aktives Material, welches eine sehr geringe Empfindlichkeit gegenüber Exzitonendissoziation aufweist, wurde eine Exzitonenkonzentration von 10^{17} /cm^3 bei einer Stromdichte von 10 kA/cm^2 berechnet. Dies ist weniger als eine Größenordnung von der Singulett-Exzitonendichte an der Laserschwelle entfernt.

Neben bimolekularen Annihilationen und der Exzitonendissoziation erschwert zusätzlich die Absorption durch im Bauteil akkumulierte Polaronen und Triplett-Exzitonen das Erreichen der Laserschwelle. Im Rahmen dieser Arbeit wurde der Einfluss von Polaronen- und Triplett-Triplett-Absorption für mehrere Bauteilgeometrien bei gleichzeitigem Wirken von Annihilationen und Exzitonendissoziation durch numerische Simulation untersucht. Sowohl für die Polaronen- als auch für die Triplett-Triplett-Absorption wurden Grenzwerte für die Absorptionswirkungsquerschnitte berechnet, bei deren Überschreiten Lasertätigkeit für alle Stromdichten verhindert wird. Bei einer Stromdichte von 10 kA/cm^2 muss der Wirkungsquerschnitt für die Triplett-Triplett-Absorption kleiner als das 5×10^{-4} -fache und der Wirkungsquerschnitt für die Polaronenabsorption kleiner als das 3×10^{-6}-fache des Wirkungsquerschnittes der stimulierten Emission sein.

Der Einfluss von Triplett-Exzitonen kann prinzipiell durch Impulsbetrieb verringert werden. Bei kurzen Anregungspulsen erreicht die Dichte der Singulett-Exzitonen ihren maximalen Wert früher als die Dichte der Triplett-Exzitonen. Dadurch könnte der Einfluss der Triplett-Triplett-Absorption um den Faktor 60 verringert werden.

Der Einfluss von Polaronen könnte prinzipiell durch Rückwärtspulse verringert werden. Hierbei wird direkt nach einem Impuls in Vorwärtsrichtung und, sobald die Singulett-Exzitonendichte ihr Maximum erreicht hat, ein kurzer Impuls in Rückwärtsrichtung angelegt. Hierbei sollte der Impuls in Rückwärtsrichtung eine etwa fünffach höhere Spannung aufweisen als der Impuls in Vorwärtsrichtung. Durch den Vorwärtsimpuls werden Ladungsträger in das Bauteil injiziert, welche in der Rekombinationszone Exzitonen bilden. Sobald die maximale Dichte an Singulett-Exzitonen erreicht ist, wird ein Impuls in Rückwärtsrichtung angelegt, welcher überschüssige Ladungsträger wieder aus dem Bauteil entfernt. Die Exzitonen verbleiben aufgrund ihres Dipolcharakters und ihrer hohen Bindungsenergie im Bauteil. Durch Simulation konnte gezeigt werden, dass Rückwärtspulse tatsächlich zu einer Reduzierung der Singulett-Exzitonendichte führen. Allerdings gestaltet sich die technische Umsetzung als sehr anspruchsvoll, da Anstiegszeiten der Impulse von etwa 1 ns und Spannungen von einigen kV erforderlich sind.

Im Rahmen dieser Arbeit wurde anhand von numerischen Simulationsrechnungen aufgezeigt, dass die Realisierung einer organischen Laserdiode in OLED-Geometrie mit den verfügbaren Materialien aktuell nicht möglich erscheint. Für Lasertätigkeit sind Singulett-Exzitonendichten von etwa 3×10^{17} /cm^3 erforderlich, was Stromdichten von einigen 1 kA/cm^2 erfordert. Durch die geringe Leitfähigkeit organischer Halbleiter treten jedoch zahlreiche Verlustprozesse in den Vordergrund, welche die Quanteneffizienz um einige Größenordnungen verringert, wodurch die erforderliche Stromdichte abermals angehoben wird. Obwohl in den letzten Jahren bedeutende Fortschritte hin zur organischen Laserdiode erzielt wurden, erscheint die Realisierung eines elektrisch gepumpten organischen Lasers in der OLED-Geometrie mit aktuell verfügbaren Materialien daher als noch nicht möglich. Es wurde darüber hinaus aufgezeigt, dass mit zukünftigen verbesserten organischen Halbleiter-Materialien sowie durch Anregung im Impulsbetrieb der Einfluss der Verlustprozesse signifikant reduziert werden könnte. Dadurch wäre es möglich, für Stromdichten im Bereich von $10\,\mathrm{kA/cm^2}$ optischen Gewinn zu erzielen. Im Rahmen dieser Arbeit wurden Materialeigenschaften und Bauteilgeometrien ermittelt, welche dann Lasertätigkeit erlauben würden. Diese Erkenntnisse können als Wegweiser für die Entwicklung neuer organischer Halbleiter für Laserdioden dienen.

Publications

Journal publications of results presented in this work

1. **C. Gärtner** and U. Lemmer, "The impact of field-induced exciton dissociation in organic laser diodes", submitted to IEEE Journal of Quantum Electronics (2009)

2. C.Pflumm, **C. Gärtner**, and U. Lemmer, "A numerical scheme to model current and voltage excitation of organic light-emitting diodes", IEEE Journal of Quantum Electronics **44** 8, pp. 790 - 798 (2008)

3. M. Punke, S.Valouch, S.W. Kettlitz, N. Christ, **C. Gärtner**, M. Gerken, and U. Lemmer, "Dynamic characterization of organic bulk heterojunction photodetectors", Applied Physics Letters **91**, 071118 (2007)

4. **C. Gärtner**, C. Karnutsch, C. Pflumm, and U. Lemmer, "Numerical device simulation of double heterostructure organic laser diodes including current induced absorption processes", IEEE Journal of Quantum Electronics **43** 11, pp. 1006-1017 (2007)

5. C. Karnutsch, C. Pflumm, G. Heliotis, J.C. deMello, D.D.C. Bradley, J. Wang, T. Weimann, V. Haug, **C. Gärtner**, and U. Lemmer, "Improved organic semiconductor lasers based on a mixed-order distributed feedback resonator design", Applied Physics Letters **90**, 131104 (2007)

6. **C. Gärtner**, C. Karnutsch, C. Pflumm, and U. Lemmer, "The influence of annihilation processes on the threshold current density of organic laser diodes", Journal of Applied Physics **101**, 023107 (2007)

7. C. Karnutsch, **C. Gärtner**, V. Haug, U. Lemmer, T. Farrell, B.S. Nehls, U. Scherf, J. Wang, T. Weimann, G. Heliotis, C. Pflumm, J.C. deMello, and D.D.C. Bradley, "Low threshold blue conjugated polymer lasers with first- and second-order distributed feedback", Applied Physics Letters **89**, 201108 (2006)

Conference proceedings and presentations of results presented in this work

1. **C. Gärtner**, C. Karnutsch, J. Brückner, N. Christ, S. Uebe, U. Lemmer, P. Görrn, T. Rabe, T. Riedl, and W. Kowalsky "Loss processes in organic double-heterostructure laser diodes", *Proceedings of the SPIE: Organic Light-Emitting Materials and Devices X, Vol. 6655*, 665525, San Diego, CA, USA (2007)

2. **C. Gärtner**, C. Karnutsch, J. Brückner, T. Woggon, S. Uebe, N. Christ, and U. Lemmer "Numerical study of annihilation processes, excited state absorption and field quenching for various organic laser diode design concepts", *9th European Conference on Molecular Electronics (ECME)*, Metz, France (2007)

3. **C. Gärtner**, Y. Nazirizadeh, C. Karnutsch, S. Uebe, and U. Lemmer, "Rate coefficients for bimolecular singlet exciton annihilation processes in organic semiconductor materials", *7*[th] *International Conference on Optical Probes of π-Conjugated Polymers and Functional Self Assemblies*, Turku, Finland (2007)

4. **C. Gärtner**, C. Karnutsch, N. Christ, and U. Lemmer, "Reducing the impact of charge carrier induced absorption in organic double heterostructure laser diodes", *Conference on Lasers and Electro-Optics (CLEO) Europe*, Munich, Germany (2007)

5. C. Pflumm, **C. Gärtner**, C. Karnutsch, U. Lemmer, "Influence of electronic properties on the threshold behavior of organic laser diode structures", *Proceedings of the SPIE: Organic Light-Emitting Materials and Devices X, Vol. 6333*, 63330W, San Diego, CA, USA (2006)

6. C. Karnutsch, V. Haug, **C. Gärtner**, U. Lemmer, T. Farrell, B.S. Nehls, U. Scherf, J. Wang, T. Weimann, G. Heliotis, C. Pflumm, J.C. deMello, and D.D.C. Bradley, "Low threshold blue conjugated polymer DFB lasers", *Conference on Lasers and Electro-Optics (CLEO)*, CFJ3, Long Beach, CA, USA, (2006)

7. **C. Gärtner**, C. Pflumm, C. Karnutsch, V. Haug, and U. Lemmer, "Numerical study of annihilation processes in electrically pumped organic semiconductor laser diodes", *Proceedings of the SPIE: Organic Light-Emitting Materials and Devices X, Vol. 6333*, 63331J, San Diego, CA, USA (2006)

8. C. Karnutsch , V. Haug, **C. Gärtner**, U. Lemmer, C. Pflumm, G. Heliotis, J.C. deMello, D.D.C. Bradley, J. Wang, and T. Weimann, "Low-threshold blue vertically emitting polyfluorene DFB lasers employing first-order feedback", *SPIE Optics + Photonics*, San Diego, CA, USA (2006)

Contents

1 Introduction   **1**
1.1 Aims and objectives 2
1.2 Structure 2

2 An introduction to organic semiconductor laser diodes   **5**
2.1 Concepts for organic semiconductors 5
     2.1.1 Electron configuration of carbon 6
     2.1.2 Conjugated molecules 6
     2.1.3 π-conjugated polymers 8
     2.1.4 Small molecules 9
     2.1.5 Conjugated dendrimers 10
     2.1.6 Summary 10
2.2 Optical properties 12
2.3 Electrical properties 14
     2.3.1 Charge carriers in organic semiconductors 15
     2.3.2 Charge transport 15
     2.3.3 Langevin recombination 16
     2.3.4 Injection of charge carriers 19
     2.3.5 Summary 23
2.4 Energy transfer 23
     2.4.1 Förster transfer 23
     2.4.2 Dexter transfer 24
     2.4.3 Summary 25
2.5 Loss processes in organic semiconductors under high excitation 26
     2.5.1 Bimolecular annihilation processes 27
         2.5.1.1 Singlet-singlet annihilation (SSA) 27
         2.5.1.2 Singlet polaron annihilation (SPA) 29
         2.5.1.3 Singlet-triplet annihilation (STA) 30
         2.5.1.4 Triplet-polaron annihilation (TPA) 30
         2.5.1.5 Triplet-triplet annihilation (TTA) 30
         2.5.1.6 Intersystem crossing (ISC) 31
         2.5.1.7 Summary 31
     2.5.2 Induced absorption processes 31
         2.5.2.1 Polaron absorption 32
         2.5.2.2 Triplet absorption 33
         2.5.2.3 Summary 33

2.5.3   Field-induced exciton dissociation 34
      2.5.3.1   Field quenching in polymers 34
      2.5.3.2   Field quenching in small molecule materials 35
      2.5.3.3   Summary 35
2.5.4   Summary 36
2.6   Organic semiconductor lasers 36
  2.6.1   Optical gain in conjugated materials 36
  2.6.2   Laser structures 37
  2.6.3   Summary 39
2.7   Concepts for organic laser diodes 39
  2.7.1   Double heterostructure organic LEDs 40
  2.7.2   Double heterostructure organic field-effect transistors 41
  2.7.3   Organic LED with field-enhanced electron transport 43
  2.7.4   Exciton accumulation 44
  2.7.5   Summary 47
2.8   Chapter summary 47

3   Modelling of organic laser diodes   49
3.1   Device geometry 49
3.2   Modelling of the optical properties 50
  3.2.1   Wave guiding 51
  3.2.2   Transfer-matrix method 52
  3.2.3   Photon density 54
3.3   Electrical modelling 55
  3.3.1   Rate equations for particles 55
  3.3.2   Drift-diffusion model 56
  3.3.3   Charge carrier injection 56
  3.3.4   Exciton generation 57
  3.3.5   Radiative and nonradiative decay 57
  3.3.6   Stimulated emission 57
  3.3.7   Laser threshold current density 58
  3.3.8   Material parameters 58
3.4   Modelling of loss processes under high excitation 60
  3.4.1   Bimolecular annihilations 60
      3.4.1.1   Rate coefficients 60
      3.4.1.2   Singlet exciton rate equation for BA 62
      3.4.1.3   Triplet exciton rate equation for BA 62
      3.4.1.4   Summary 63
  3.4.2   Field-induced exciton dissociation 63
      3.4.2.1   1D Poole-Frenkel theory 64
      3.4.2.2   3D Poole-Frenkel theory 65
      3.4.2.3   Onsager theory 65
      3.4.2.4   Hopping separation model 67
      3.4.2.5   Material parameters 67

		3.4.2.6	Summary	68
	3.4.3		Induced absorption processes	68
		3.4.3.1	Net modal gain	68
		3.4.3.2	Gain and attenuation coefficients	69
		3.4.3.3	Definition of absorption parameters	70
	3.4.4		Summary	71
3.5			Chapter summary	71

4 Impact of bimolecular annihilation processes — 73

4.1		Device geometry of organic TE_0-mode laser diode	73
4.2		Impact of annihilation processes on the laser threshold current density	75
4.3		Bimolecular annihilations	76
	4.3.1	Singlet-singlet annihilation (SSA)	78
	4.3.2	Singlet-triplet annihilation (STA)	78
	4.3.3	Singlet-polaron annihilation (SPA)	81
	4.3.4	Triplet-triplet annihilation (TTA)	81
	4.3.5	Triplet-polaron annihilation (TPA)	82
	4.3.6	The role of triplet excitons	83
4.4		Chapter summary	84

5 Impact of polaron and triplet-triplet absorption — 87

5.1			Polaron absorption and bimolecular annihilations	87
5.2			Polaron and triplet absorption parameters	89
5.3			Laser threshold current density	92
5.4			Impact of device geometry	94
	5.4.1		Emission layer thickness	94
	5.4.2		Transport layer thickness	96
5.5			Impact of material properties	98
	5.5.1		Charge carrier mobilities	98
		5.5.1.1	Transport layer	98
		5.5.1.2	Emission layer	99
		5.5.1.3	Emission and transport layer	100
		5.5.1.4	Field dependent mobilities	101
	5.5.2		Variation of exciton lifetimes	102
		5.5.2.1	Singlet exciton lifetime	102
		5.5.2.2	Triplet exciton lifetime	103
5.6			Pulsed operation	104
	5.6.1		Transient absorption parameters	104
	5.6.2		Switch-on characteristics of an organic laser diode	105
	5.6.3		Charge carrier mobility	106
	5.6.4		Triplet quenchers	108
	5.6.5		Reverse current pulses	108
	5.6.6		Summary	110
5.7			Chapter summary	110

6 Impact of field-induced exciton dissociation   **113**
6.1 Device geometry 113
6.2 Threshold current density 114
     6.2.1 Bimolecular annihilation processes 114
     6.2.2 Waveguide attenuation 115
     6.2.3 Field quenching 117
6.3 Impact of device geometry 118
     6.3.1 Emission layer thickness 120
     6.3.2 Transport layer thickness 120
6.4 Impact of material properties 121
     6.4.1 Charge carrier mobilities 121
     6.4.2 Exciton binding energies 123
     6.4.3 Annihilation processes 125
6.5 Implications for induced absorption processes 126
6.6 Guest-Host-Systems 127
6.7 Chapter summary 128

7 Design rules for organic laser diodes   **131**

8 Conclusions and outlook   **133**

9 Acknowledgement   **135**

Bibliography   **147**

Index   **167**

1 Introduction

Organic optoelectronic devices such as organic solar cells (OSCs), light emitting diodes (OLEDs), organic photodiodes (OPDs) and organic field-effect transistors (OFETs) have been employed successfully for more than one decade [1–6]. However, the realization of an organic laser diode still remains unaccomplished. Since almost 10 years, there have been vigorous attempts to realize optical gain under intense electrical excitation. Organic semiconductor laser diodes (OLDs) are expected to provide a number of unique features such as the tunability over the whole visible range and low-cost, large area fabrication. These properties would render OLDs suitable for a huge number of applications in bioanalytics, digital printing, fluorescence spectroscopy and homeland security.

In fact, there have been two reports of electrically excited optical gain in organic materials [7, 8]. The first report [7] was discredited due to scientific misconduct [9]. In the second paper, sharp emission at 410 nm from an Alq$_3$:DCM device was reported [8]. In fact, pristine Alq$_3$ emits in the green and DCM is a red dye. The spectrally narrow blue emission has been attributed to a sharp atomic emission line from indium, since an indium electrode had been employed. Recently, there have been two reports of spectrally narrowed edge emission from electrically excited organic devices [10,11]. However, the authors couldn't demonstrate any threshold behavior as a function of the excitation density in order to clearly separate waveguide effects from stimulated emission. The authors of the first paper attributed the observed spectral line narrowing to waveguide cutoff and misalignment of the optical setup [10].

Optically pumped organic semiconductor lasers have been shown as early as 1996 [12–14]. In addition, low-threshold laser operation has been demonstrated [15–18]. Recently, a breakthrough with hybrid laser devices has been achieved, where an organic semiconductor laser is pumped with an inorganic laser diode [14,19–22] or an inorganic light emitting diode [23] allowing low-cost integrated tunable laser devices. The attempt to pump an organic semiconductor laser with an organic light emitting diode was so far unsuccessful [24]. Although high current densities have been realized for thin-film devices [25–30], neither an electrically pumped organic laser nor an electrically pumped organic amplifier has been shown yet.

This thesis analyzes the impact of various loss processes on the threshold current density of organic semiconductor laser diodes by numerical simulation. The electrical properties of the device are described by a self consistent drift-diffusion model. Electrical processes are included employing material parameters of typical small-molecule organic semiconductor materials. The following processes are taken into account: field-dependent injection of charge carriers, drift and diffusion of particles, Langevin recombination, stimulated and spontaneous emission, waveguide attenuation, absorption by polarons and triplet excitons, field-induced dissociation of excitons as well as annihilation processes between charge carriers, singlet and triplet excitons. The optical properties are calculated using a transfer matrix method. Design concepts based on organic double-heterostructures are evaluated which provide good exciton confinement and exhibit a low-attenuation waveguide. Parameter studies are performed in order to identify crucial material and device parameters whose optimization is promising. Design rules are derived which can be used in order to reduce the impact of various loss processes and in order to improve the device performance. The question is addressed, whether an organic injection laser can be realized for the slab organic double-heterostructure design, if material properties in the range of typical OLED materials are assumed.

1.1 Aims and objectives

In organic semiconductors, optical gain is generated by molecules in the first excited singlet state which are referred to as singlet excitons. Under electrical excitation with current densities above 10 A/cm^2, various loss processes become important which efficiently reduce the singlet exciton density. Hence, the optical gain is diminished by orders of magnitude. Charged molecules (polarons) and molecules in the first triplet state both quench singlet excitons and also induce additional absorption losses, whereby the optical gain is again decreased.

The objective of this thesis is to study the device physics of organic double-heterostructure laser diodes and to clarify, if laser operation can be achieved in organic materials by electrical pumping. The impact of various loss mechanisms is to be modeled and their impact on the threshold current density shall be studied. In addition, different device concepts shall be evaluated regarding their sensitivity to induced loss mechanisms. Design rules shall be formulated, which can be employed in order to reduce the impact of loss processes.

1.2 Structure

In the first chapter, the topic of the thesis is defined and explained. In addition, the structure of this work is outlined.

In the second chapter, the basics for this thesis are derived. First, an introduction to organic semiconductors is given and loss processes under high excitations are introduced. Subsequently, a short review on recent developments in the field of optically pumped organic lasers is given. Finally, alternative concepts for electrically pumped organic laser diodes are presented. The chapter is concluded with a brief summary.

In the third chapter, the modelling of organic laser diodes is presented. First, the modelling of the waveguide using a transfer matrix method is discussed. Then, the drift-diffusion model is elucidated, which is used for describing electrical properties of organic laser devices, and the coupling of both models is specified.

In the fourth chapter, simulation results for organic laser diodes are discussed which have been obtained using the simulation models described in the third chapter. The laser threshold current density is calculated and the impact of bimolecular annihilation processes on the laser threshold is discussed.

In the fifth chapter, the impact of polaron and triplet-triplet absorption on the performance of the considered devices is analyzed. In addition, the transient behavior of the optical gain and the absorption parameters is studied.

In the sixth chapter, simulation results are presented which analyze the impact of field-induced dissociation of singlet and triplet excitons on the laser threshold current density.

In the seventh chapter, design rules for organic laser diodes are briefly summarized, which can be used in order to reduce the impact of loss processes on the threshold current density significantly.

In the eighth chapter, the results of this thesis are summarized and an outlook on further developments is given.

2 An introduction to organic semiconductor laser diodes

Lasers based on organic semiconductor materials offer mechanical flexibility, low-cost production and tunability at the same time. Available organic lasers are pumped optically by additional laser sources such as inorganic solid state lasers or inorganic deep-blue laser diodes and light emitting diodes. Electrical pumping of an organic laser diode is expected to provide a further substantial reduction of the system costs, since an additional inorganic laser diode is not needed, which would open new applications for organic lasers in bioanalytics and spectroscopy. Despite of recent advances in stability and reduction of laser threshold under optical pumping, an organic laser diode has not been realized yet. One important issue is that interaction of basic processes are still not understood.

This chapter gives an introduction into the fundamentals of organic laser diodes. In section 2.1, the basic concepts for organic semiconductors are described and the underlying physical mechanisms are explained. The optical and electronic properties of organic semiconductors are described in section 2.2 and section 2.3. Energy transfer in organic materials is described in section 2.4. Section 2.5 discusses various loss processes which especially occur at high excitation and whose impact is in particular striking for organic laser diodes. Design concepts for optically pumped organic lasers and the possibility of indirect electrical pumping are discussed in section 2.6. Possible concepts for electrically pumped organic laser diodes are described in section 2.7. A short summary of this chapter is given in section 2.8.

2.1 Concepts for organic semiconductors

Organic compounds are chemical compounds which contain a covalently bound carbon atom. Due to the large number of binding configurations of carbon, the number of known organic materials is huge. Today, more than 19 million different organic compounds with various properties are known. The optical and electrical properties of these molecules can be designed by altering their molecular structure. Therefore, organic materials are of great interest for optoelectronic application. Since the discovery of conductive polymers [31], organic materials with semiconducting and metallic characteristics are available.

In this section, the electron configuration of the carbon atom is discussed which is the reason for the huge variety of organic materials. Following this, the concept of conjugated molecules is elucidated. After this, three classes of organic semiconductors, small molecules, conjugated polymers and conjugated dendrimers, are presented and aspects concerning their processing technique are discussed.

2.1.1 Electron configuration of carbon

Carbon atoms exhibit in total 6 electrons. For unbound carbon atoms in the ground state, the electrons are arranged in the $1s^2 2s^2 2p^2$ configuration, which is shown schematically in figure 2.1 (top left). For the ground state, the 1s- and the 2s-orbitals are occupied by two electrons and two out of three 2p-orbitals are occupied by one electron. When carbon atoms are bound, hybrid orbitals are formed by the 2s- and the three 2p-Orbitals.

In figure 2.2, the structures of ethane, ethene and ethyne are shown which are examples of different bond configurations of carbon. For ethane, the carbon atoms form four single bonds, which are arranged at the corners of a tetrahedron and the included angle between two bonds is 109.5°. In this case, the carbon atoms exhibit a sp^3-hybridization, whose electron configuration is shown in figure 2.1, top right. These bonds are qualified as σ-bonds, since they exhibit a cylindrical symmetry. The sp^3-hybridization is the common configuration of the carbon atom for materials exhibiting single bonds. These bonds are also called saturated bonds.

For ethene, the carbon atom exhibits the sp^2-hybridization. The electron configuration for the sp^2-hybridization is shown schematically in figure 2.1, bottom left: 3 sp^2-orbitals and one p-orbital are formed by the s-orbital and three p-orbitals. The carbon atoms are linked by a double bond, which consists of a strongly bound σ-bond of the sp^2-orbitals and a weakly bound π-bond, which is formed by the p_z-orbitals of the carbon atoms.

Besides single and double bonds, carbon is also able to form triple bonds. Figure 2.2, right shows the structure of ethyne where the two carbon atoms are linked by a triple bond. In this case, the carbon atoms exhibit the sp-hybridization, whose electron configuration is shown in figure 2.1, bottom right. In this case, the carbon atoms are bound by σ-bond of the sp-orbitals and two weakly bound π-bonds of the p-orbitals.

2.1.2 Conjugated molecules

For organic compounds exhibiting exclusively single bonds, four σ-bonds per carbon atom are formed (sp^3-hybridization). The binding energy of a single bond is above 3 eV. At room temperature, these bonds are in the ground state and no charge carriers are available for electrical conduction. Hence, these compounds are electrically isolating. Popular examples for single bound organic compounds are organic polymers such as polyethene, polystyrene, polypropene or polyvinyl chloride which are often employed as cladding of electric cables. Single bound organic compounds appear transparent for visible light and are absorbing for wavelengths of the far ultraviolet light below 160 nm (corresponding to a photon energy of 7.7 eV). Since the

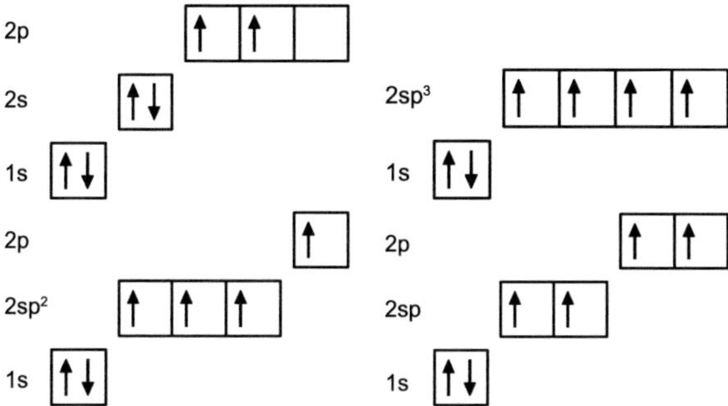

Figure 2.1: Electron configuration of carbon: ground state (top left), sp^3- (top right), sp^2- (down left) and sp-hybridization (down right). For the ground state, the 1s- and the 2s- orbitals are occupied by two electrons and two 2p-orbitals are occupied by one electron. Hybrid orbitals are formed by superposition of s- and p-orbitals.

Figure 2.2: For ethane, carbon has four single bonds (sp^3-hybridization). For ethene, the two carbon atoms are bound by a double bond (sp^2-hybridization). Here, the 2p-orbitals form a π-bond. For ethyne, carbon is even able to form a triple bond (sp-hybridization) which consists of a σ-bond and two π-bonds.

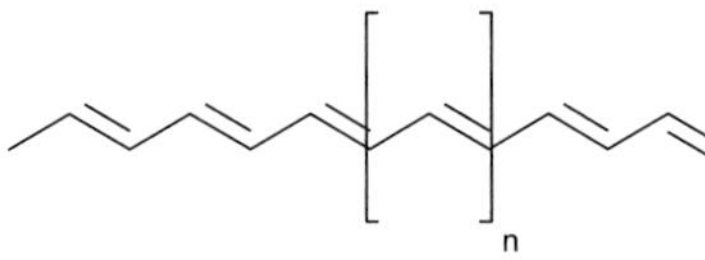

Figure 2.3: Chemical structure of polyethyne which is a basic example for a conjugated polymer. The electrons of the π-bond are weakly bound and delocalized along the polymer backbone. The π-electrons can be excited by visible light and electric charge transport is possible.

excitation energy is above the dissociation energy, saturated hydrocarbons cannot be employed as active materials in optoelectronic devices.

Molecules exhibiting alternating single and double bonds are referred to as conjugated molecules. For conjugated molecules, the π-electron is delocalized. In figure 2.3, the chemical structure of the polymer polyethyne is shown, which is a basic example for a conjugated compound. This polymer exhibits alternating single and double bonds. The carbon atoms exhibit the sp^2-hybridization and form three strongly bound σ-bonds and one weakly bound π-bond. While the σ-bonds form the skeletal structure of the molecule the π-electrons (which form the π-bonds) are delocalized along the polymer chain. Due to the valence of carbon, single and double bonds are alternating along the polymer backbone. The repeating structural unit is indicated by two brackets.

The binding energy of the π-electron is around 1 eV. By electrical excitation or irradiation with visible light, the π-electron can be excited to the anti-binding π^*-state. The energy gap between the binding π- and the anti-binding π^*-state constitutes a bandgap, which is typically around 2.5 eV. Additionally, charge transport is enabled along the polymer chain. Thus, conjugated molecules exhibit the features of a semiconductor.

Organic semiconductors can be classified by the type of processing. Devices can be fabricated from π-conjugated polymers by solution processing. Another class of organic semiconductors are small-molecule materials, which exhibit small molecular weight, and which can be processed by thermal evaporation in high vacuum. These two types of materials are presented in the next sections. A third class of organic semiconductors, conjugated dendrimers, is briefly discussed in section 2.1.5.

2.1.3 π-conjugated polymers

In figure 2.4, the chemical structures of some frequently used *π-conjugated polymers* is shown. Organic semiconducting polymers can be processed from solution by spin-coating, various printing techniques and doctor blading. Thus, the production technology is low-cost and large area production is feasible. The fabrication

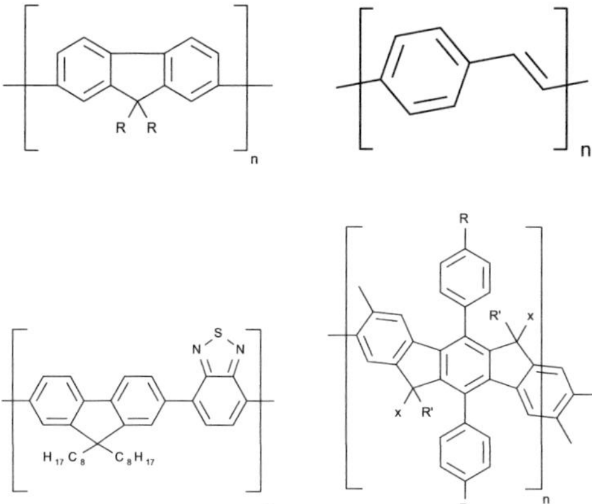

Figure 2.4: Chemical structure of conjugated polymer materials: poly(9,9-dioctylfluorene) (PFO) (top left), polyphenylene vinylene (PPV) (top right), poly(9,9-dioctylfluorene-co-benzothiadiazole (F8BT) (down left) and ladder-type polyparaphenylene (MeLPPP) (down right)

of multilayer devices from solution is challenging, since the upper layers may not dissolve the subjacent layers. Hence, materials with orthogonal solvents have to be employed, which greatly reduces the combination of materials, which can be used in a device. Alternatively, cross-linkable polymers are available which become insoluble after illumination.

Electroluminescence in conjugated polymers has been demonstrated in 1990 by Burroughes et. al. [32]. Optical pumped laser operation has been shown in 1996 by Tessler and Hide [12, 33].

2.1.4 Small molecules

Besides polymers, conjugated molecules with low-molecular weight can also be used for fabrication of organic optoelectronic devices. These materials are referred to as *small molecule*-materials which also exhibit systems of delocalized π-electrons. Due to their low molecular weight, these compounds can be processed by thermal evaporation under high vacuum. Hence, arbitrary combinations of materials and layers can be fabricated. By co-evaporation, these materials can be doped in order to improve the conductivity and in order to tune the emission characteristics of these materials. Optically induced laser operation in small molecule-semiconductors has been shown in 1997 by Kozlov and Berggren [34, 35].

In figure 2.5, the chemical structures of several small-molecule materials are shown.

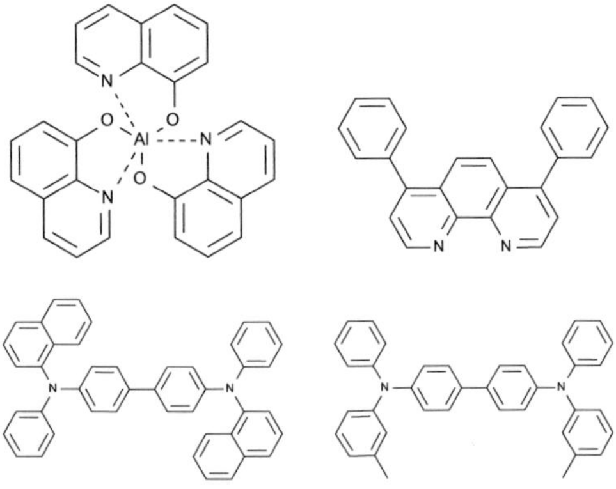

Figure 2.5: Chemical structures for the small molecule materials aluminum tris(8-hydroxyquinoline) (Alq$_3$), 4,7-Diphenyl1,10-phenanthroline (Bphen), N,N'-di(naphthalene-1-yl)N,N'-diphenylbenzidine (NPB) and N,N'-DiphenylN,N'-bis(m-tolyl)benzidin (TPD). These materials are used in organic light-emitting diodes (OLEDs) and organic lasers.

2.1.5 Conjugated dendrimers

Conjugated dendrimers [36–44] are a class of molecules whose molecular weight is between small molecules and conjugated polymers. The word *dendrimer* has its roots in the Greek words for tree (*dendros*) and parts (*meros*). In figure 2.6, the structure of a dendrimer is shown schematically. Conjugated dendrimers are based on a modular design principle and consist of *conjugated units* (C) and *surface groups* (S), which are connected by *linkers* (L). These materials exhibit a fractal like structure and their growing is based on self-assembly. Thus, a high degree of complexity is achieved. Dendrimers have been employed successfully in OLEDs and in organic solar cells [45, 46].

2.1.6 Summary

In this section, the basic concept for organic semiconductors has been discussed. Conjugated molecules exhibit alternating single and double bonds enabling a delocalization of the π-electrons along the molecule. Due to their low binding energy, π-electrons can be excited electrically and optically by photons in the visible range. Thus, conjugated molecules offer semiconducting features like their inorganic counterparts.

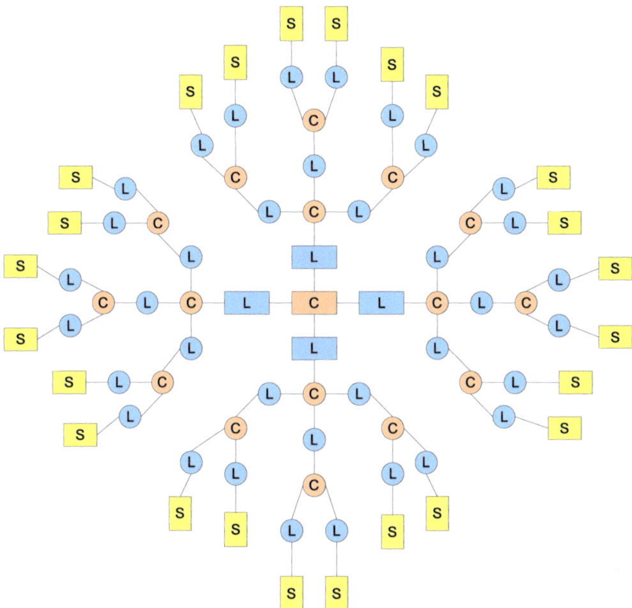

Figure 2.6: Conjugated dendrimer (schematically) consisting of conjugated units (C), linkers (L) and surface groups (S). The optical properties of the molecular core can be tuned independently from the electrical properties of the surrounding branches.

2.2 Optical properties

The semiconducting nature of conjugated molecules is constituted by the energy difference of around 2.5 eV between the binding π- and the antibinding π^*-state. The π-bonds can be excited much more easily than the strongly bound σ-bonds. Hence, for a description of the semiconducting properties, the electrons of the σ-bonds can be neglected. For the ground state, the highest occupied molecular orbital (HOMO) can be identified with the π-state. The lowest unoccupied molecular orbital (LUMO) of the molecule can be identified as the antibinding π^*-state. By absorption of a photon, an electron-hole pair is generated which is delocalized along the molecule. The Coulomb binding energy of the charge pair is

$$E = \frac{e^2}{4\pi\epsilon_0\epsilon_r d}. \tag{2.1}$$

Here, e is the elementary charge, ϵ_0 is the dielectric constant of vacuum, ϵ_r is the dielectric material constant and d is the separation distance of the charge carriers. The dielectric material constants ϵ_r of organic semiconductors are relatively small when compared to their inorganic counterparts (gallium arsenide: $\epsilon_r = 13$, PPV: $\epsilon_r \approx 3$). Thus, the coulomb attraction of charge pairs in organic semiconductors is considerably stronger than in inorganic semiconductors. Hence, the separation distance d becomes smaller for organic materials and the exciton exhibits a higher binding energy. The binding energies of the charge pairs are of the order of 1 eV which is large compared to the thermal activation energy $k_\mathrm{B}T = 25$ meV. Hence, excitons are stable at room temperature and the excitonic nature of charge pairs is pronounced in organic materials. The activation scheme of organic molecules is therefore treated within the exciton image. The ground state of the molecule is associated with the lowest excitonic state S_0.

Due to vibronic degrees of freedom, the excitonic states are split into various vibronic substates. In thermal equilibrium, most molecules are in the lowest vibronic states. The energy difference between the ground state S_0 and the first excited singlet state S_1 is referred to as bandgap.

In figure 2.7, the transition intensities of a molecule during absorption and emission are shown schematically. Organic semiconductors are absorbing for photon energies above the bandgap energy. Molecules absorbing a photon undergo a transition from the S_0 state to a vibronic subband of the S_1 state. The excited molecule quickly relaxes from higher vibronic states to the lowest vibronic state of the S_1 band with a typical lifetime of 1 ps. Radiative and nonradiative decay may occur from the lowest vibronic subband of the S_1 band to the excitonic ground state with a lifetime in the range from 10^{-10} s to 10^{-8} s. In this case, transitions are allowed, whose energy is equal or smaller than the bandgap energy.

When a molecule absorbs light, the electrical charges within the molecule rearrange. Thus, the Coulomb forces within the molecule as well as the normalized nucleus coordinate r are changed. While optical transitions occur on a timescale of 10^{-15} s, the rearrangement of the nuclei is much slower (typically 10^{-12} s) [47]. In a good

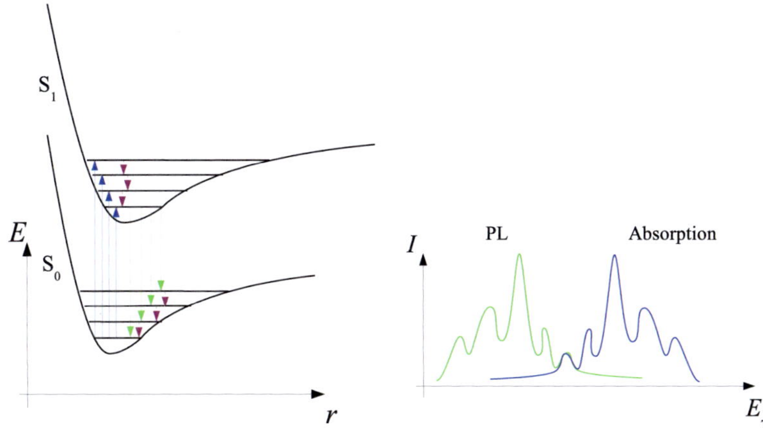

Figure 2.7: Energy diagram of the excitonic ground state S_0 and the first excited singlet state S_1. Under optical excitation, organic semiconductors are absorbing for photon energies above the bandgap energy. The absorbing molecule undergoes a transition from the ground state S_0 to a vibronic substate of the S_1 band. Following this, the molecule thermalizes with a lifetime of 1 ps to the lowest vibronic state. From the S_1 state, the molecule decays radiatively with a typical lifetime of 1 ns. Due to the relaxation mechanism, the photoluminescence (PL) is red-shifted relatively to the absorption (Stokes-shift).

approximation, optical transitions occur for fixed atomic nuclei. Thus, according to the Born-Oppenheimer approximation [47], optical transitions occur *vertically*. The probability of coupling with vibronic modes for optical transitions are described by the Franck-Condon factors [47].

When single molecules are examined, sharp absorption and emission maxima are observed, which correspond to radiative transitions between the vibronic states. Due to the different transition rules for absorption and emission, the absorption spectra are red shifted relatively to the emission spectra and the emission and absorption spectra exhibit a mirror symmetry. The energy difference between the maximum of the absorption and emission spectra is referred to as Stokes shift. For laser devices, materials exhibiting a large Stokes shift are favorable due to a reduced self absorption.

Figure 2.8 shows the Jablonski diagram of an organic molecule [47]. After absorption of a photon, the molecule is excited to a singlet state. For high enough photon energies, the exciton can also be excited to the higher lying singlet states S_2 ... S_N. After relaxation, excitons can undergo fluorescent decay. Alternatively, the exciton can undergo intersystem crossing from a singlet to a triplet state. This transition

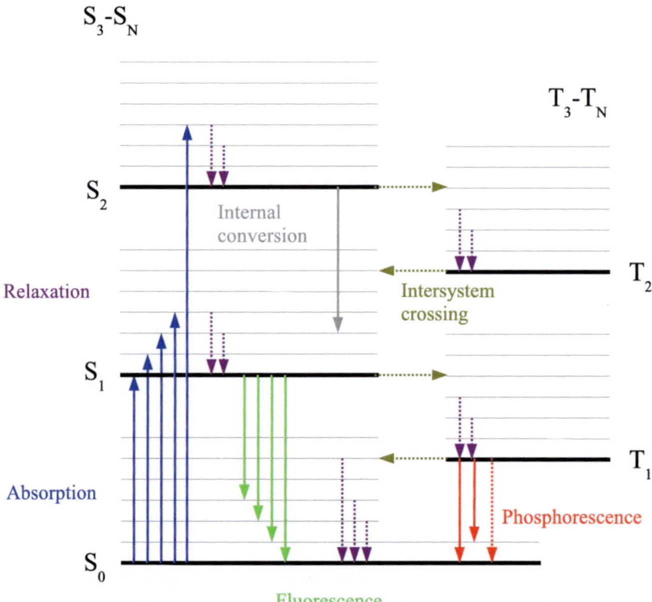

Figure 2.8: Jablonski-diagram of an organic molecule showing the basic transitions between the ground state S_0 and the excited singlet and triplet states S_1, S_2, T_1 and T_2. For details see text.

is quantum mechanically forbidden, since it comprises a spin flip. Thus, for typical fluorescent OLED materials the lifetimes for intersystem crossing are generally much higher than the lifetimes for fluorescent decay. For phosphorescent organic semiconductors, which exhibit a metal atom with a strong spin-orbit coupling, this transition becomes enhanced.

For fluorescent molecules, excitons in the T_1-state exhibit a lifetime in the range from 10^{-5} s up to 10^{-1} s. For triplet excitons, phosphorescent decay or inverse intersystem crossing are possible deactivation mechanisms. Here, the nonradiative decay is dominating. The radiative decay is increased for specific molecules comprising a metal atom.

2.3 Electrical properties

In this section, the electrical properties of organic semiconductors are discussed. First, a general overview of charge carriers in organic materials is given. Then, the injection of charge carriers from a metal electrode and the transport of charges within the organic layers is discussed. Following this, the recombination of oppositely charged particles to excitons by Langevin recombination is elucidated. Finally, the topics of this section are summarized.

2.3.1 Charge carriers in organic semiconductors

Due to the low dielectric constant of organic semiconductors, the interaction between charge carriers and fixed atoms of the molecule is much stronger than in inorganics. Charge carriers induce a lattice distortion within their surrounding area. Hence, the interaction of charge carriers with molecular vibrations (phonons) is by far stronger pronounced than in inorganic semiconductors. Therefore, the properties of charge carriers in organic semiconductors in combination with lattice distortions are usually described as a quasiparticle, which is referred to as *polaron*. Due to relaxation of the molecule, the polaronic states are within the bandgap between the HOMO and LUMO levels. Due to the additional states within the bandgap, the presence of polarons can be determined experimentally by sub-bandgap absorption of organic semiconductors.

Figure 2.9 schematically sketches the energy levels for positively and negatively charged polarons. For positive polarons P^+ (also referred to as hole or radical cation), the lower polaronic state is occupied by only one charge. Hence, the total charge of the molecule is positive. For negatively charged polarons P^- (also referred to as electron or radical anion), the lower polaronic level is occupied with two charges, which are filled from charges in the HOMO level. The upper level is occupied by an additional single charge.

For organic laser diodes, these polaronic states induce additional absorption bands. Polaron induced absorption is assumed to be an important loss mechanism and a major obstacle towards the realization of an organic laser diode. Therefore, the impact of polaron absorption in organic laser diodes has to be considered.

2.3.2 Charge transport

As mentioned previously, charge carriers are delocalized along the conjugated molecule. Thus, additional charges can move freely along the conjugation length. However, most polymer and small molecule materials do not exhibit a long-range translational symmetry and the delocalization of charges is not extended over several molecules. Thus, charges are localized at a segment of a polymer chain or the considered molecule. Charge transport in organic semiconductors can be understood as a stochastic hopping process between different sites. Charge transfer between different molecules is based on phonon assisted tunneling, which is also referred to as *hopping transport*. While intrachain mobilities of charges (within the molecule) of nearly 600 cm^2/Vs have been reported [48], the overall mobility is greatly limited by charge hopping. For actual organic semiconductors, the mobility is in the range from

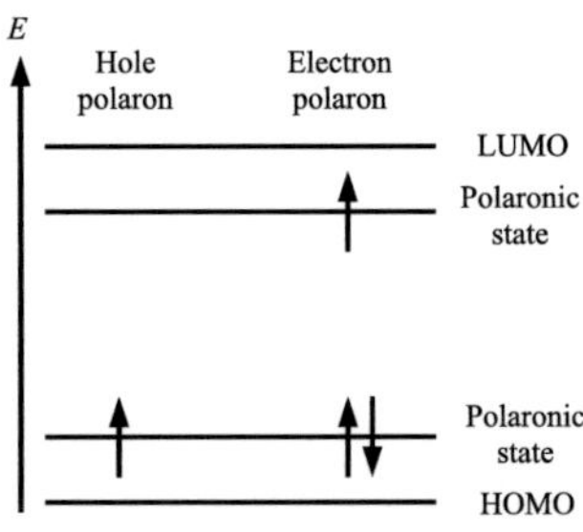

Figure 2.9: Schematic energy diagram of polaronic levels. New polaronic states are formed within the bandgap. In organic laser diodes, additional induced absorptions are caused by polarons.

10^{-8} cm^2/Vs to 10^{-3} cm^2/Vs. Hence, hopping is the mechanism, which limits the mobility in organic semiconductors.

Transport in organic semiconductors has been described theoretically by the disorder formalism by H. Bässler [49–51]. Figure 2.10 sketches the mechanism of charge transport in organic semiconductors for electrons following the Bässler model. These states are distributed around the HOMO and LUMO levels following a Gaussian distribution. Bässler proposed, that transport occurs by hopping between localized states. The actual mobility depends on the energetic distribution of states and the distribution of the hopping distances. For this model, good agreement has been achieved with experimental data for the temperature and field dependence of the mobility.

An alternative model for charge transport in organic semiconductors is the Poole-Frenkel approach, which describes charge transport as the field-assisted escape of charge carriers from a lower lying state, and which is employed in our model. The field and temperature dependent mobility is given by

$$\mu(F,T) = \mu_0 \exp\left(-\frac{(F/F_0)^{1/2}}{k_\mathrm{B}T}\right). \tag{2.2}$$

For the field- and temperature dependent mobility in actual devices, good agreement is achieved for the Poole-Frenkel theory.

2.3.3 Langevin recombination

For inorganic semiconductor devices, light emission is achieved, when electrons and holes meet within the semiconductor and recombine radiatively. For organic semiconductors, the mechanism for light generation is different. Due to the low dielectric constant of organic materials, excitons are stable at room temperature. Thus,

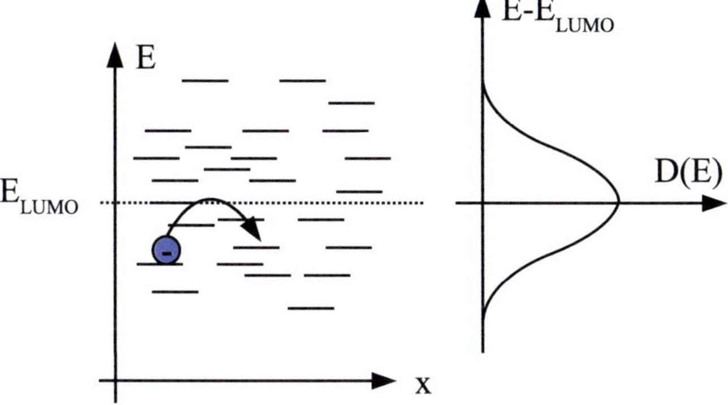

Figure 2.10: Charge transport in organic semiconductors is based on hopping between localized states. The mobility is determined by the distribution of distances between sites and their energetic distribution.

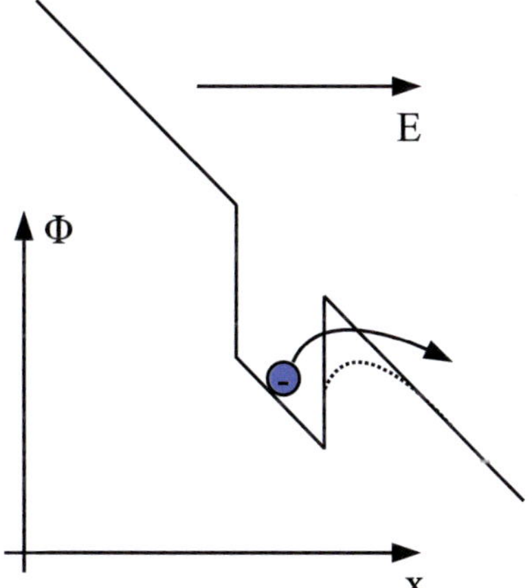

Figure 2.11: Within the Poole-Frenkel model, charge transport is modeled by field-induced escape of charges from lower lying states below the conduction band, which are modeled as coulomb wells.

emission is not achieved by direct recombination of opposite charge carriers. In organic semiconductors, excitons are generated by recombination of electrons and holes. Light emission is then achieved from radiative decay of excitons.

Within the Langevin theory [52], the dynamics of charges is described, whose motion is governed rather by random diffusion than by drift. When oppositely charged carriers are within their Coulomb radius

$$r_{\mathrm{c}} = \frac{e}{4\pi\epsilon_0\epsilon_{\mathrm{r}}k_{\mathrm{B}}T} \tag{2.3}$$

they are attracted to each other by their electric field. Since their thermalization energy is smaller than their Coulomb binding energy, the charge carriers will subsequently undergo Langevin recombination, whereby an exciton is generated. The reaction for exciton formation is

$$e + h \rightarrow \begin{cases} {}^1\!/\!_4: & \mathrm{S}_{\mathrm{N}}^* \rightarrow \mathrm{S}_1 \\ {}^3\!/\!_4: & \mathrm{T}_{\mathrm{N}}^* \rightarrow \mathrm{T}_1 \end{cases} \tag{2.4}$$

Here, two oppositely charged carriers recombine and an exciton in a higher lying singlet state $\mathrm{S}_{\mathrm{N}}^*$ or triplet state $\mathrm{T}_{\mathrm{N}}^*$ is formed. The exciton relaxes from higher excitonic states by interconversion within about 1 ps to the lowest excited state S_1 or T_1. During interconversion, the spin multiplicity of the exciton is conserved. Due to spin statistics, three times more triplet excitons than singlet excitons are generated, which has also been verified experimentally [53]. For organic laser diodes, only the fluorescent singlet exciton state is able to generate optical gain. Additionally, triplet excitons exhibit additional absorption bands, which cause additional attenuation of the laser mode in organic laser diodes. Hence, triplet excitons are generally undesirable in organic laser diodes.

The exciton formation rate by Langevin recombination is

$$k_{\mathrm{Langevin}} = \frac{e\left(\mu_{\mathrm{e}} + \mu_{\mathrm{h}}\right) n_{\mathrm{e}}\left(x, t\right) n_{\mathrm{h}}\left(x, t\right)}{\varepsilon_r \varepsilon_0}. \tag{2.5}$$

In organic lasers, gain is generated by singlet excitons. Since Langevin recombination is the main process for singlet exciton generation, the actual value is of crucial importance. As shown in equation (2.5), k_{Langevin} is a function of the charge carrier mobilities μ_{e} and μ_{h}, indicating that high mobility materials are favorable for a high exciton generation rate. For many organic semiconductors, the electron mobility is about one order of magnitude smaller than the hole mobility. In this case, the Langevin rate is determined by the higher mobility. On the other hand, k_{Langevin} is a function of the product of the electron and hole densities $n_{\mathrm{e}}\left(x, t\right) \times n_{\mathrm{h}}\left(x, t\right)$. Hence, the Langevin recombination rate is strongly increased for higher particle densities.

2.3.4 Injection of charge carriers

For electrical excitation of organic semiconductor devices, conductive electrodes are required for charge injection. In OLEDs, transparent conductive oxides (TCO) and metals are employed as electrode materials. While typical operation current densities of OLEDs are of the order of some mA/cm^2, much higher current densities are necessary in organic laser diodes. It is assumed that current densities above 100 A/cm^2 are needed for laser operation. Therefore, the electrodes have to enable high current densities which depend on excellent injection properties.

Current injection into an organic device can occur in two different regimes. For large injection barriers between the metal and the organic layer, the current is limited by injection (injection limited current (ILC)). The maximum current density is achieved for space-charge limited current (SCLC). Here, the impact of the injection barrier is negligible and the current density is only limited by the space charge within the organic device.

Besides good injection, the optical properties of the electrodes are also of crucial importance. As it turned out for optically pumped devices, the attenuation coefficient of the waveguide should be below 5 cm^{-1}. The optical properties of the electrodes have a huge impact on the attenuation coefficient. Typical attenuation coefficients for metals are of the order of 10^6 cm^{-1} and around $5{\times}10^3$ cm^{-1} for typical TCOs. Hence, either the strongly reduced attenuation coefficients for the contact materials or a small filling factor of the laser mode within the contacts is a prerequisite in order to achieve low waveguide attenuation. Thus, suitable electrode materials have to combine good injection with low waveguide absorption.

In this section, the physics of charge injection from a conductive electrode into an organic semiconductor is discussed. Usually, two models are used for describing carrier injection into organic semiconductors: firstly, carrier injection by thermionic emission or secondly, by Fowler-Nordheim tunneling. The model by Scott and Malliaras [54], which is employed in this work, treats injection as a hopping process in the presence of an electric field and recombination at the electrode.

Thermionic emission The thermionic emission model considers the lowering of the image charge potential by the external electric field [55]. This model has also been employed for inorganic semiconductors by Schottky [55], describing the temperature dependent injection properties of a junction between a metal and an inorganic crystalline semiconductor. Figure 2.12 sketches the interface between a metal and an inorganic semiconductor. E_F is the Fermi level of the metal, Φ_A is the work function, Φ_{LUMO} is the LUMO level of the organic semiconductor and $\Phi_B = \Phi_A - \Phi_{LUMO}$ is the height of the injection barrier.

For the thermionic emission model, the current density is given as a function of the field

$$J_{TE} = aT^2 \exp\left(-\frac{\Phi_B - b\sqrt{F}}{k_B T}\right). \qquad (2.6)$$

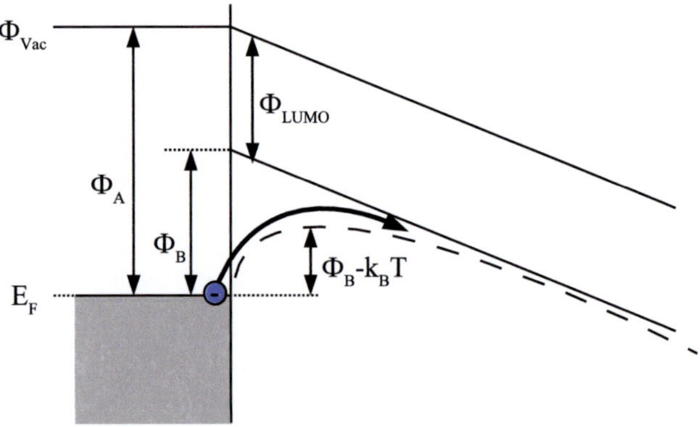

Figure 2.12: Carrier injection from a metal contact into an organic semiconductor.

The parameters a and b are defined as follows

$$A^* = 4\pi e m^* k_{\mathrm{B}}^2 h^{-3},$$

(2.7)

$$b = \sqrt{e^3/4\pi\epsilon\epsilon_0}.$$

(2.8)

Here, A^* is called the *effective Richardson constant* [55]. For inorganic semiconductors, the calculations have been carried out with the assumption that charges propagate freely in the conduction band. For highly amorphous organic semiconductors however, this assumption is not justified, since transport is based on hopping between localized sites. Additionally, the thermionic emission model is not applicable to low mobility semiconductors. Due to the large concentration of charge carriers at the interface, backflow of charges will occur [56]. Improved models have been developed for low-mobility semiconductors and insulators [56,57].

Fowler-Nordheim tunneling Alternatively, emission from a metal electrode into a semiconductor can be described in terms of field emission, also called Fowler-Nordheim (FN) tunneling. Here, charge carriers are transferred from the electrode into the insulator by quantum mechanical tunneling. This model has been employed successfully for describing charge carrier injection from metals into thick insulators like silicon oxide. Figure 2.13 sketches the emission of an electron into an organic semiconductor by tunneling through a triangular potential barrier.

The current density as a function of the electric field is

$$J_{\mathrm{FN}} = \frac{A^* e^2 F^2}{\Phi_{\mathrm{B}} C^2 k_{\mathrm{B}}^2} \exp\left(-\frac{2C\Phi_B^{1.5}}{3eF}\right),$$

(2.9)

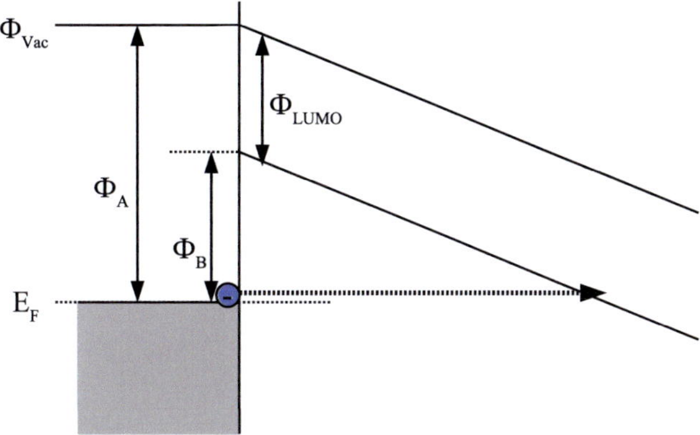

Figure 2.13: Carrier injection from a metal contact into an organic semiconductor by Fowler-Nordheim tunneling.

where

$$C = \frac{4\pi\sqrt{2m^*}}{h},\qquad(2.10)$$

and A^* has been defined in equation 2.7, and m^* is the reduced charge carrier mass.

FN tunneling assumes that charges can move freely within the conductance band of the semiconductor or insulator. For organic semiconductors, this picture may not apply, since transport is dominated by hopping. Additionally, organic semiconductors do not exhibit a sharp band edge but a smooth distribution of states for the HOMO and LUMO levels. Hence, additional states within the triangular potential barrier and a random distribution of the effective barrier potentials may be present. These effects have been studied analytically [58] and by Monte-Carlo simulations.

Scott-Malliaras model The models of thermionic emission and Fowler-Nordheim tunneling, which have been discussed in the previous paragraphs, do not take into account that charge transport is rather dominated by hopping between localized states than by free propagation within extended bands [49, 57].

In our calculations, we have employed a different model, which has been presented by Scott and Malliaras [54] in 1998. This model treats charge injection from a metal electrode into an organic semiconductor as a hopping process in the presence of an electric field and the Coulomb potential of the charge pair. Additionally, the Langevin recombination near the metal surface of the injected carrier and its image charge is considered (see section 2.3.3).

Within the model of Scott and Malliaras, the image charge potential is considered, which is lowered by the applied electric field. As derived for the thermionic emission model, the potential within the semiconductor is

$$\Phi\left(x\right) = \Phi_{\mathrm{B}} - eFx - \frac{e^2}{16\pi\epsilon_0\epsilon_r x}. \tag{2.11}$$

The injection barrier is denoted as Φ_{B}. In the case of Langevin recombination, the injected electron and its image charge will definitely recombine, if the coulomb binding energy $E_{\mathrm{C}}\left(x\right)$ exceeds $k_{\mathrm{B}}T$. This is the case within the Coulomb radius r_{C}. For the recombination current, following equation is derived:

$$J_{\mathrm{rec}} = n_0 e \mu E\left(x_{\mathrm{C}}\right) = \frac{16\pi\epsilon_0\epsilon_r k_{\mathrm{B}}^2 T^2 n_0 \mu}{e^2}. \tag{2.12}$$

Here, n_0 denotes the charge density at the interface and μ is the electron mobility. For zero field, the total current is zero:

$$J_{\mathrm{inc}} - J_{\mathrm{rec}} = C\left[\exp\left(-\frac{\Phi_{\mathrm{B}}}{k_{\mathrm{B}}T}\right) - \frac{n_0}{N_0}\right] = 0, \tag{2.13}$$

with

$$C = \frac{16\pi\epsilon_0\epsilon_r k_{\mathrm{B}}^2 T^2 N_0}{e^2} = A^* T^2. \tag{2.14}$$

The density of chargeable sites is denoted N_0 and the Richardson constant is denoted A^*. By considering the field dependence of recombination and injection, the field dependent net current is obtained:

$$J_{\mathrm{SM}} = 4\Psi^2 N_0 e \mu F \exp\left(-\frac{\Phi_{\mathrm{B}}}{k_{\mathrm{B}}T}\right) \exp f^{0.5}, \tag{2.15}$$

where the reduced field f is defined

$$f = \frac{eFr_{\mathrm{C}}}{k_{\mathrm{B}}T}, \tag{2.16}$$

and

$$\Psi\left(f\right) = f^{-1} + f^{-0.5} - f^{-1}\left(1 + 2f^{0.5}\right)^{0.5}. \tag{2.17}$$

The derived expression includes thermionic injection and carrier back-flow into the contact due to the low mobilities. The above considerations are also valid for hole injection into the HOMO, when the respective values are identified. In our model, we employ the Scott-Malliaras model for current densities, which are directed away from the contact. Currents, which are directed towards the electrodes, remain unaffected.

2.3.5 Summary

In this section, we have summarized the electrical properties of organic semiconductors. In contrast to their inorganic counterparts, transport is rather based on hopping between localized states than on free propagation within delocalized bands. Models describing properties for organic semiconductors have to take this fact into account. For charge injection, the model by Scott and Malliaras is employed, which considers thermionic emission and recombination at the electrode. In our model, the field dependent mobility is described by a Poole-Frenkel model. When oppositely charged carriers encounter, excitons are formed by Langevin recombination. Due to their high binding energy, excitons are stable at room temperature. For all employed models, good agreement with experimental results is obtained.

2.4 Energy transfer

Besides charge transport, organic molecules exhibit the ability to exchange energy and charges on a distance of several nanometers. These energy transfer mechanisms are especially important for guest-host systems and under high excitation. The most trivial energy transfer mechanism is radiative emission and reabsorption by a different molecule. In organic semiconductors, two further energy transfer mechanisms are present, which will be discussed in this section.

2.4.1 Förster transfer

Fluorescent resonant energy transfer (FRET), also called Förster transfer, is a resonant dipole-dipole transfer between singlet exciton states of two molecules. This energy transfer mechanism has been described by Förster in 1948 [59]. The mechanism of FRET is sketched in figure 2.14. An excited donor molecule (D*) is deactivated and its energy is transferred to an accepting molecule (A). Förster transfer is based on *Fermi's golden rule* for energy transfer between states. The transfer rate Γ_{mn} of a donor state m to an acceptor state n is

$$\Gamma_{mn} = 4\pi^2 \left| V_{mn} \right|^2 \delta \left(E_n - E_m \right),$$ (2.18)

where V_{mn} is the matrix element for interaction of the states m and n with energies E_m and E_n. The energies of the donating and accepting states have to match, which is stated by the δ-function. Hence, a prerequisite for Förster transfer is a spectral overlap of the emission spectrum of the donor and the absorption spectrum of the acceptor (see figure 2.15). By evaluating the interaction matrix elements and integrating over the spectrum of the donor and acceptor, the *Förster formula* is derived, which states the total transfer rate:

$$\Gamma_{DA} = 8.8 \times 10^{17} \frac{\kappa^2}{n^4 \tau_D R^6} \int g_A \left(E \right) g_D \left(E \right) dE$$ (2.19)

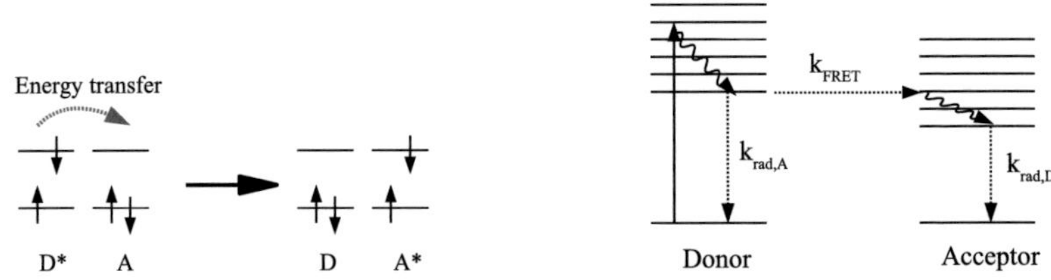

Figure 2.14: Schematic illustration of Förster energy transfer. Left: An excited molecule (donor) is deactivated and its energy is transferred to an accepting molecule. Right: The absorbing energies of the acceptor have to match the emitting energies donor.

The fluorescent lifetime of the donor is τ_D, the value of κ depends on the relative orientation of the dipoles, n is the refractive index and R is the geometric distance of donor and acceptor. The normalized spectra for absorption of the acceptor and for fluorescence of the donor are $g_A(E)$ and $g_D(E)$. Employing the definition of the *Förster radius R_0*,

$$R_0 = \left(8.8 \times 10^{17} \frac{\kappa^2}{n^4} \int g_A(E) \, g_D(E) \, dE \right)^{1/6}, \qquad (2.20)$$

equation 2.19 can be written

$$\Gamma_{DA} = \frac{1}{\tau_D} \left(\frac{R_0}{R} \right)^6. \qquad (2.21)$$

Since Γ_{DA} is a function of R^{-6}, Förster energy transfer strongly depends on the distance between donor and acceptor. For organic materials, Förster transfer is a long range mechanism, which occurs on length scales in the range from 1 to 10 nm. Förster transfer requires that transitions are allowed for the donor and the acceptor. Hence, only transitions employing singlet excitons are allowed.

The Förster transfer mechanism is in particular important for the energy transfer in guest-host systems. Several bimolecular annihilation processes, which occur in organic materials under high excitation and whose impact is in particular striking for organic laser diodes, are based on Förster energy transfer (see section 2.5).

2.4.2 Dexter transfer

Dexter transfer is a short-range process, where an exciton diffuses from an excited donor state to an acceptor state [60]. The basic mechanism for Dexter transfer is sketched in figure 2.16. In contrast to Förster transfer, where energy is transferred by resonant dipole-dipole coupling, charge carriers are transferred simultaneously in this case. In figure 2.16, left, the transfer of a singlet exciton is shown. Here, an

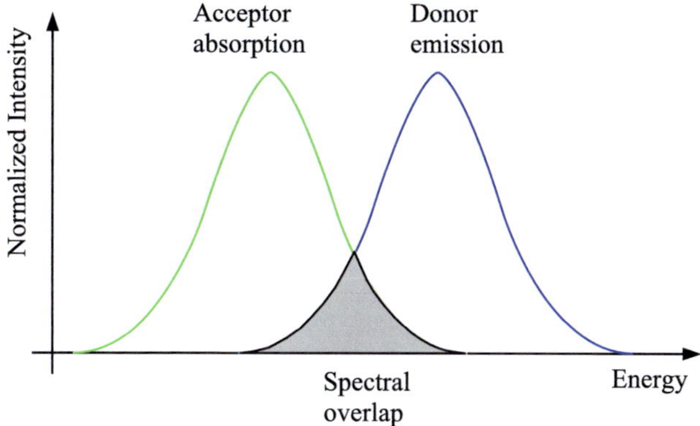

Figure 2.15: Schematic illustration of Förster energy transfer. An excited molecule (donor) is deactivated and its energy is transferred to an accepting molecule.

electron in the LUMO state of the donor is transferred to the LUMO state of the acceptor. Simultaneously, an electron from the HOMO of the acceptor is transferred to the HOMO of the donor, which corresponds to a hole transfer from the donor to the acceptor. The same mechanism can be applied to the transfer of triplet excitons, which is sketched in figure 2.16, right-hand side. The rate for Dexter transfer is

$$\Gamma_{\text{Dexter}} = \hbar P^2 J \exp\left(-\frac{2r}{L}\right), \qquad (2.22)$$

where r is the distance between the donor and the acceptor, P and L are constants (which are not easily accessible by experiment, for details see [60]) and J denotes the spectral overlap of the excitonic states of the donor and the acceptor. Since the Dexter transfer rate is a function of $\exp(-r)$, Γ_{Dexter} strongly depends on the transfer distance. Typically, Dexter transfer is a short range transfer, which only occurs on distances in the range from 0.5 nm to 2 nm.

2.4.3 Summary

In this section, energy transfer mechanisms in organic semiconductors have been discussed. Energy can be exchanged between singlet exciton states of molecules by the long-range Förster transfer, which is based on a dipole-dipole interaction. Förster transfer is the mechanism, which acts in guest-host systems. A second transfer mechanism is the Dexter transfer, where an entire exciton is delivered from the donor to the acceptor by exchange of charges. Both transfer mechanisms play an important role in organic laser diodes, where loss channels by bimolecular annihilation processes

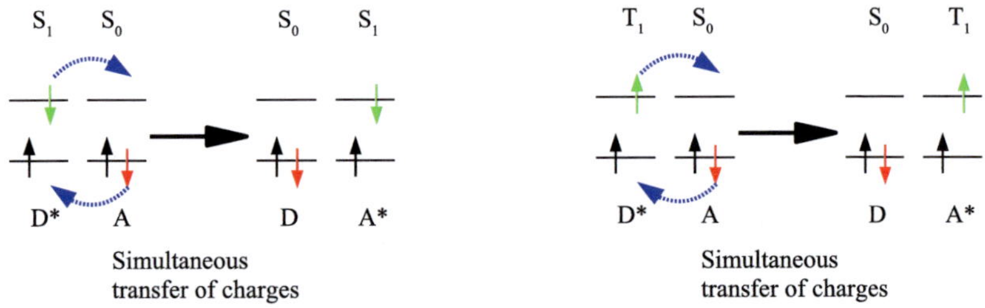

Figure 2.16: Schematic illustration of Dexter transfer. The energy is transferred by exchange of a charge pair. Thus, only the complete spin of the donor and the acceptor has to be conserved. Therefore, the transfer of singlet and triplet excitons is allowed.

occur at high excitation.

2.5 Loss processes in organic semiconductors under high excitation

Organic laser diodes are expected to exhibit threshold current densities around $1 \ \text{kA/cm}^2$. However, when compared to inorganic semiconductors, the mobility in organic semiconductors is reduced by many orders of magnitude. While inorganic materials for optoelectronic applications exhibit charge carrier mobilities above $1000 \ \text{cm}^2/\text{Vs}$, the mobility in organic semiconductors is in the range from 10^{-8} to $10^{-3} \ \text{cm}^2/\text{Vs}$. Therefore, particle densities of up to $10^{19} \ /\text{cm}^3$ and electric fields of up to $10^7 \ \text{V/cm}$ are required in order to achieve current densities of the order of $1 \ \text{kA/cm}^2$. Hence, at high current densities, various induced losses may become important, such as bimolecular annihilation processes as well as charge carrier induced absorption by polarons and triplet excitons. Due to the immense electric field, field-induced exciton dissociation is expected to be an additional loss channel.

In this section, the physical background for loss processes is given, which become important in organic semiconductors under high excitation. First, bimolecular annihilation processes are discussed in section 2.5.1. Induced absorption losses by polaron and triplet-triplet absorption are treated in section 2.5.2. Section 2.5.3 describes the impact of field-induced exciton dissociation (field quenching, FQ) in organic semiconductors and discusses theoretical models for FQ. An overview of current research on loss processes in organic semiconductors is given. The modelling of loss processes will be presented in section 3.4. This section will be concluded with a short summary in section 2.5.4.

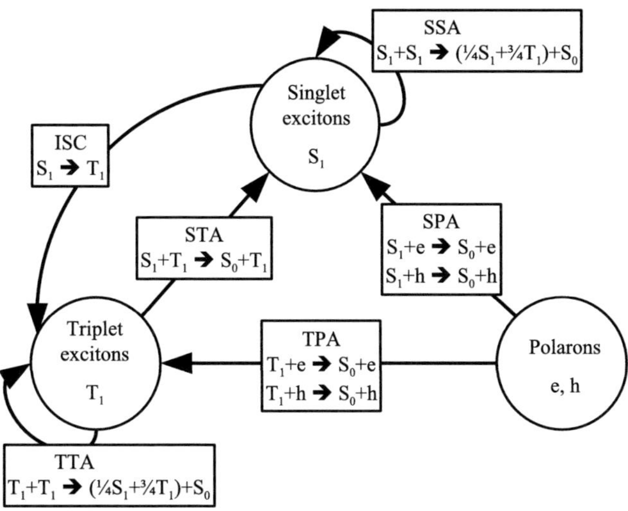

Figure 2.17: Overview of bimolecular annihilation processes in organic semiconductors. Most annihilation processes quench singlet excitons, being the species responsible for optical gain.

2.5.1 Bimolecular annihilation processes

Bimolecular annihilation processes are intermolecular processes, which occur between charge carriers and excitons. For the considered processes, two particles are involved, hence the effective annihilation rate depends on the product of two particle densities. Thus, these processes are especially important at high excitation, where high particle densities are present. Figure 2.17 provides an overview of the annihilation processes being considered in the model. The arrows indicate their main impact on the density of a particle species. Particle species are indicated by circles. Most processes have a negative impact on the density of the singlet excitons. Therefore, a significant impact on the threshold of organic laser diodes is expected. In this section, the physical background and basic mechanisms of bimolecular annihilation processes are discussed.

2.5.1.1 Singlet-singlet annihilation (SSA)

The singlet-singlet annihilation (SSA) process describes the bimolecular process, whereby two singlet excitons interact [61, 62]. When two singlet excitons collide, the energy of the first exciton is transferred to the second exciton, which is excited to a higher lying state. Subsequently, a singlet or a triplet exciton is formed. In our model, spin-statistics determine the probability of singlet and triplet exciton formation [53]. The singlet exciton ratio is described by the parameter ξ. We assume that three times more triplet than singlet excitons are generated and therefore $\xi = 0.25$ [61]. The reaction equations for SSA are shown in equation (2.23) and equation (2.24). The reaction rate coefficient is denoted κ_{SSA}, the first excited singlet state, the first excited triplet state and ground state are denoted S_1, T_1 and S_0, respectively.

$$S_1 + S_1 \xrightarrow{(1-\xi) \times \kappa_{SSA}} T_1 + S_0 \tag{2.23}$$

$$S_1 + S_1 \xrightarrow{\xi \times \kappa_{SSA}} S_1 + S_0 \tag{2.24}$$

In order to estimate the impact of bimolecular annihilation processes on the laser threshold of an actual device, the absolute values of the annihilation rate constants are of vital importance. For singlet-singlet annihilation, the rate constant κ_{SSA} can be obtained from transient photoluminescence measurements [63]. In this case, an organic semiconductor is excited by a short laser pulse exhibiting a pulse width of about 1 ps (which is much less than the typical radiative lifetime of singlet excitons of organic semiconductors). For low excitation densities, SSA is negligible, hence the singlet exciton density n_{S_1} follows the following differential equation

$$\frac{dn_{S_1}(t)}{dt} = -\frac{n_{S_1}(t)}{\tau_{S_1}}, \tag{2.25}$$

which is fulfilled by

$$n_{S_1}(t) = n_{S_1,0} \exp\left(-\frac{t}{\tau_{S_1}}\right), \tag{2.26}$$

where $n_{S_1,0}$ is the initial singlet exciton density. Since $I(t) \sim \tau_{S_1}^{-1} \times n_{S_1}(t)$, the photoluminescence intensity $I(t)$ follows an exponential law

$$I(t) = I_0 \exp\left(-\frac{t}{\tau_{S_1}}\right), \tag{2.27}$$

where I_0 is the initial photoluminescence intensity. When $\log I(t)$ is plotted as a function of time, the monomolecular lifetime can be obtained by linear fitting.

The annihilation rate coefficient κ_{SSA} can be obtained from transient photoluminescence measurements at high excitation. For excitation intensities above $10^{16}\,/\mathrm{cm}^3$, the impact of SSA becomes important. In this case, the rate equation for the transient photoluminescence is

$$\frac{dn_{S_1}(t)}{dt} = -\frac{n_{S_1}(t)}{\tau_{S_1}} - (2-\xi)\,k_{SSA}n_{S_1}^2(t), \tag{2.28}$$

where ξ is the probability of generation of an singlet exciton for SSA (here $\xi = 0.25$). The transient photoluminescence is described by the following function

$$I(t) = \frac{I_0}{\exp\left(\frac{t}{\tau_{S_1}}\right) + (2 - \xi)\, k_{SSA} n_{S_1,0} \tau_{S_1} \left(\exp\left(\frac{t}{\tau_{S_1}}\right) - 1\right)}. \tag{2.29}$$

Since the excitation pulses are much shorter than the radiative exciton lifetime of the organic semiconductor, the initial excitation density $n_{S_1,0}$ can be derived from the excitation energy density and the absorption coefficient of the considered material. Since k_{SSA} is the only free parameter, the SSA rate coefficient can be then obtained from a fit to experimental data.

Recently, an alternative method has been proposed, where the SSA rate coefficient is obtained from the efficiency roll-off in bilayer organic light emitting diodes [64]. However, under electrical excitation, singlet excitons are also quenched by further annihilation processes, where polarons and triplet excitons are involved. Hence, the obtained rate coefficient has to be corrected for the impact of quenching by polarons and triplet excitons. In addition, the leakage current increases due to charge carrier imbalance, which reduces the quantum efficiency and might partly cause the observed efficiency roll-off behavior.

2.5.1.2 Singlet-polaron annihilation (SPA)

Singlet excitons are also quenched by polarons. Singlet-polaron annihilation (SPA) is an Auger-like process, where the singlet exciton decays and its energy is transferred to the polaron, which is then excited to a higher polaronic state. Subsequently, the polaron quickly undergoes relaxation. SPA occurs for interaction of holes and electrons with singlet excitons. The rate coefficients for both SPA processes are not necessarily equal. For singlet-electron annihilation (SPA,e), the rate coefficient is $\kappa_{SPA,e}$, for singlet-hole annihilation (SPA,h), the rate coefficient is $\kappa_{SPA,h}$. The reaction equations for both SPA,e and SPA,h are:

$$S_1 + e \xrightarrow{\kappa_{SPA,e}} S_0 + e \tag{2.30}$$

$$S_1 + h \xrightarrow{\kappa_{SPA,h}} S_0 + h \tag{2.31}$$

For singlet-polaron annihilation, only a few experimental measurements are available. Here, the detailed dynamics of both singlet and triplet excitons have to be taken into account. A successful technique has been presented by List et al. [65], where experimental results from steady-state photoinduced absorption (PA), photoluminescence (PL), PL-detected magnetic resonance (PLDMR), and PA-detected magnetic resonance have been combined. The obtained data has been analyzed employing a rate equation model, which describes the dynamics of singlet and triplet excitons.

2.5.1.3 Singlet-triplet annihilation (STA)

Singlet excitons can also be quenched by triplet excitons via singlet-triplet annihilation (STA). Due to the spin-forbidden nature of phosphorescence, the radiative lifetime of triplet excitons is several orders of magnitude higher than the radiative singlet exciton lifetime. Therefore, triplet excitons use to accumulate in the device. Thus, the annihilation by triplet excitons poses an additional relevant loss channel for singlet excitons.

In this process, the energy of a singlet exciton is transferred to a triplet exciton, which is excited to a higher lying excited triplet state. Subsequently, the excited triplet exciton quickly relaxes by intercombination to the first excited triplet state. The rate coefficient is denoted by κ_{STA}. The equation is

$$S_1 + T_1 \xrightarrow{\kappa_{\mathrm{STA}}} T_1 + S_0. \tag{2.32}$$

2.5.1.4 Triplet-polaron annihilation (TPA)

Triplet excitons are quenched by polarons. Similar to singlet-polaron annihilation, triplet-polaron annihilation (TPA) is an other Auger-like process. In this case, the energy of the triplet exciton is transferred to the polaron, which is excited to a higher lying state. The rate coefficient for TPA is κ_{TPA}. The reaction equations for TPA are

$$T_1 + e \xrightarrow{\kappa_{\mathrm{TPA}}} S_0 + e \tag{2.33}$$

$$T_1 + h \xrightarrow{\kappa_{\mathrm{TPA}}} S_0 + h \tag{2.34}$$

The rate coefficient of TPA has been measured by probing the change of the phosphorescent lifetime under applied voltage in monopolar devices [66].

2.5.1.5 Triplet-triplet annihilation (TTA)

In our model, triplet-triplet annihilation (TTA) is also considered. The reaction equations for this process are shown in equation (2.35) and equation (2.36): two triplet excitons interact and the energy of the first exciton is transferred to the second exciton, which is excited to a higher lying state. It quickly relaxes either to the first excited singlet or triplet state. The average ratio of singlet excitons produced per reaction is again given by ξ. As for the TTA process, we assume that three times more triplet than singlet excitons are generated, hence $\xi = 0.25$. The rate coefficient is denoted by κ_{TTA}.

$$T_1 + T_1 \xrightarrow{(1-\xi)\times\kappa_{\mathrm{TTA}}} T_1 + S_0 \tag{2.35}$$

$$T_1 + T_1 \xrightarrow{\xi\times\kappa_{\mathrm{TTA}}} S_1 + S_0 \tag{2.36}$$

2.5.1.6 Intersystem crossing (ISC)

Our model also includes intersystem crossing (ISC), which describes the radiationless transition of singlet into triplet excitons induced by spin-orbit coupling. In organic materials, the lowest vibronic state of the first singlet band overlaps with higher vibronic states of the first triplet band. Hence, in materials with strong spin-orbit coupling, singlet excitons can efficiently be transformed into higher vibronic states of the first triplet band. Subsequently, the triplet exciton relaxes to the lowest triplet state. The reaction equation describing ISC is shown in equation (2.37) where the rate coefficient for ISC is denoted with κ_{ISC}.

$$S_1 \xrightarrow{\kappa_{\text{ISC}}} T_1 \tag{2.37}$$

Reverse intersystem crossing (RISC) describes the process whereby an exciton is transformed from the triplet state into a lower lying singlet state. Additionally, thermally excited triplet excitons occupying a higher vibronic state of the T_1 band can cross into the first excited singlet state S_1. The absolute rates for reverse intersystem crossing depend on the energy difference of the two states, and they are typically very small since higher vibronic states are usually unoccupied. Therefore, reverse intersystem crossing is not included in the presented model.

2.5.1.7 Summary

In this section, annihilation processes between polarons and excitons have been discussed. Most annihilation processes reduce the density of singlet excitons, which is the species responsible for optical gain. Therefore, the quantum efficiency is greatly reduced at high excitation densities. In order to certainly estimate the impact of annihilation processes for an actual material, the complete set of rate coefficients is needed. In table 3.3, some published rate coefficients for bimolecular annihilation processes in different polymer and small molecule materials are summarized. Thus, additional experimental investigations of annihilation processes are needed and more rate coefficients have to be determined.

2.5.2 Induced absorption processes

In the last section, the impact of polarons and triplet excitons on the singlet exciton density by bimolecular annihilation processes has been described. Additionally, polarons as well as triplet excitons induce additional absorption bands. If the laser wavelength is within an polaronic or triplet absorption band, the laser mode is attenuated. The attenuation coefficients for these induced absorption losses depend on the absorption cross section (which is a function of the wavelength), but also on the modal intensity profile and the density distribution of polarons, singlet and triplet excitons within the device. For polaron absorption, a photon of the laser mode is absorbed by a charge carrier in the HOMO or LUMO level. The charge carrier is excited to a higher polaronic state, from where it quickly relaxes back to thermal

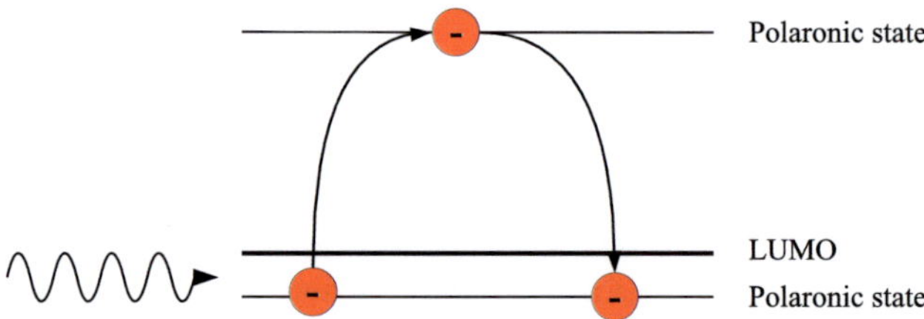

Figure 2.18: Schematic illustration of polaron absorption in organic semiconductor materials. For details see text.

equilibrium. For triplet absorption, a triplet exciton is excited to a higher triplet state. Subsequently, the excited triplet exciton relaxed by intercombination back to the lowest triplet exciton level. In this section, the physical fundamentals of polaron and triplet absorption are discussed and a review of experimental work on this topic is given.

2.5.2.1 Polaron absorption

Polaron absorption has been extensively analyzed by photoinduced absorption spectroscopy (PA) for organic semiconductors, where charges are added to a molecule by doping, charge injection or by photogeneration [67]. Polaron absorption can be understood as a subband transitions of polarons which is illustrated schematically in figure 2.18. Two polaronic species can be distinguished, namely single charged polarons and double charged bipolarons. Single charged polarons are paramagnetic particles ($S = 1/2$). While the occupation of states would allow three transitions for polarons [68–71], only two transitions are allowed due to an alternating parity of the polaronic levels [72–76]. For bipolarons, which are diamagnetic particles ($S = 0$), only one optical subbandgap transition (instead of two) is observed.

Polaron absorption has been studied for conjugated polymers [26,77–90] and small molecular weight materials [89,91,92]. For poly(2-methoxy-5-(3',7'-dimethyl)-octyloxyp-phenylenevinylene) (OC1C10-PPV), a polaron absorption cross section of 7.5×10^{-17} cm^2 has been measured [93]. For poly(p-phenylenevinylene) (PPV), a polaron absorption cross section of 10^{-15} cm^2 is assumed [26], which is about one order of magnitude higher than typical cross sections for stimulated emission [30,83,94–98]. For PPV, maximum polaron absorption is achieved at a photon energy of 1.6 eV ($\lambda = 780$ nm) [26]. However, the absorption spectrum is still significant for higher energies. Hence, organic laser diodes operating at visible wavelengths would probably be affected by polaron absorption. For Alq$_3$:DCM, electrically pumped devices are affected by polaron absorption for wavelengths in the range from 620 to 705 nm [91].

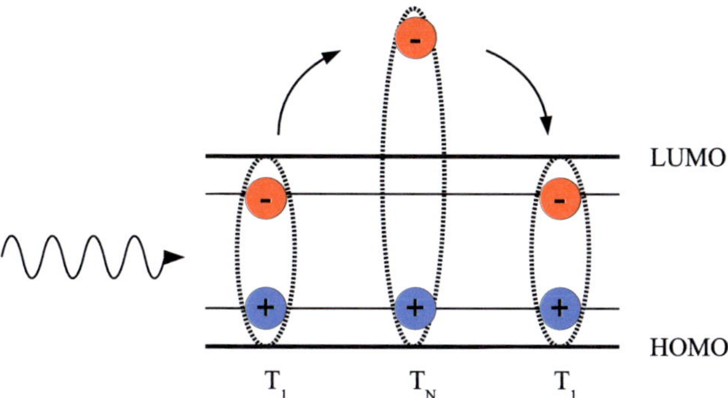

Figure 2.19: Schematic illustration of triplet absorption in organic semiconductor materials. For details see text.

2.5.2.2 Triplet absorption

Singlet and triplet excitons can also quench laser light whereby the excitons are excited to a higher lying state. The excitation of an singlet exciton into a higher lying state can be taken into consideration by using the effective cross section for stimulated emission [99]. For actual laser materials, the effective cross section for stimulated emission is still positive, hence laser operation can be achieved. Since triplet excitons do not generate optical gain, the impact of triplet absorption has to be treated in a more elaborate manner.

The triplet absorption process is sketched schematically in figure 2.19. If a photon is absorbed by an exciton in the first excited triplet state T_1, the triplet exciton is excited to a higher lying triplet state T_N. Subsequently, the excited triplet exciton quickly relaxes by interconversion to lower lying states. Triplet absorption in the near infrared has been studied by optical spectroscopy. For poly(2-methoxy-5-(3',7'-dimethyl)-octyloxyp- phenylenevinylene) (OC1C10-PPV), a triplet absorption cross sections in the range from 3×10^{-15} cm^2 to 1.6×10^{-15} cm^2 have been obtained by measurement [100–102] and coupled-cluster quantum-chemical calculations [93, 103, 104]. For F8BT, a triplet absorption cross section of 3.1×10^{-16} cm^2 was found [105]. Thus, for the analyzed conjugated polymer materials, the triplet absorption cross section is of the same order of magnitude as the stimulated emission cross section.

2.5.2.3 Summary

In this section, induced absorption losses from polarons and triplet excitons have been discussed. In organic materials under electrical excitation, polarons as well as triplet excitons are generated. Since both particles exhibit absorption bands in the near infrared and the visible range, the optical gain of an organic injection laser can be

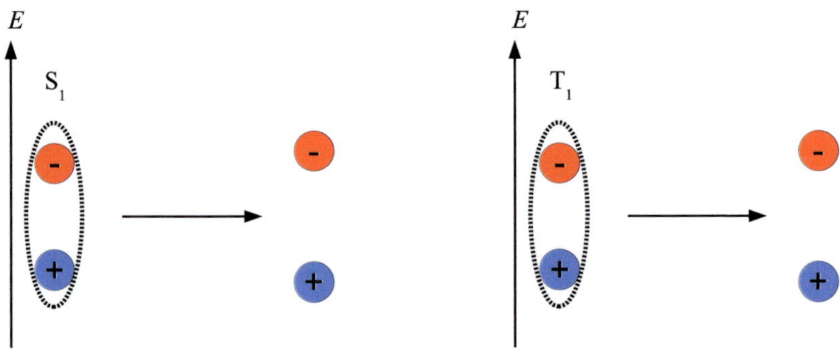

Figure 2.20: Schematic illustration of field-induced dissociation of singlet and triplet excitons.

reduced. The absorption bands exhibit a spectral width of several 10 nm. Therefore, it might be possible to separate stimulated emission from polaron and triplet absorption, spectrally.

2.5.3 Field-induced exciton dissociation

In organic semiconductors, electric fields of the order of 10^7 V/cm have to be applied in order to achieve current densities of the order of 1 kA/cm^2, which is expected to be necessary in order to achieve electrically pumped laser operation. Experimental studies for small molecule and polymer materials have shown that the photoluminescence efficiency is reduced by up to 90% in the presence of an electric field [106,107]. The basic process of field-induced dissociation of singlet and triplet excitons is illustrated in figure 2.20. Here, excitons are assumed to be separated to unbound charge carriers. Therefore, these loss processes have to be taken into account in organic laser diodes.

In this section, the basic mechanism of field-induced exciton dissociation in polymer and small molecule materials, also referred to as field quenching (FQ), is considered and some experimental results are discussed.

2.5.3.1 Field quenching in polymers

In organic semiconductors, the singlet exciton binding energy is in the range of 0.2 eV to about 1.0 eV whereas triplet exciton binding energies are reported in the range from 0.4 eV to about 1.6 eV [108–119]. Hence, excitons are stable at room temperature. In the absence of an electric field, the dominating deactivation channels are radiative and nonradiative decay, while the exciton dissociation rate to free charge carriers is rather small. When an electric field is applied, the exciton dissociation rate is greatly increased. Depending on the exciton binding energies, field induced exciton dissociation becomes an important loss mechanism for electric fields above 10^6 V/cm [107]. Here, lower dissociation rates require higher exciton binding energies.

Exciton dissociation in amorphous materials is commonly described by the Onsager theory [120], which has been extended by Braun [121] and Goliber [122]. For this theory, satisfying agreement with measurements has been achieved [123–127]. However, the need for considering the dynamics of exciton relaxation has been demonstrated [117, 128, 129], suggesting that exciton dissociation might not be described solely by a rate equation. Besides the Onsager theory, further models describing the impact of FQ have been discussed in literature [130, 131]. In previous studies, the Poole-Frenkel theory and a hopping separation theory have been employed [132]. These theories will be discussed in more detail in section 3.4.2. However, the Onsager theory seems to describe the effect of FQ correctly for most materials [125, 130, 131].

Experimental values for field-induced photoluminescence quenching are only available up to electric fields of up to 4 MV/cm [107, 133]. However, for the considered device and material properties, it turns out in our work that local fields up to 20 MV/cm would be necessary in order to achieve laser threshold. We note that this is well above the electrical breakdown field of typical thin film organic semiconductor devices.

2.5.3.2 Field quenching in small molecule materials

In small molecule materials, the impact of field-induced exciton dissociation has also been demonstrated. For pristine Alq$_3$ films, quenching efficiencies of up to 50% has been observed [130, 133]. In principal, the dissociation mechanisms in conjugated polymers and small molecule materials are comparable. However, while conjugated polymers are essentially one dimensional systems, small molecule materials have a two or even three dimensional character. Additionally, excitons in small molecule materials are more localized, therefore, the binding energy is higher and smaller dissociation rates are expected.

Luo et. al. [133] studied the impact of field quenching in dye-doped Alq$_3$ films. When compared to pristine Alq$_3$, the quenching efficiency could be reduced dramatically for small-bandgap dyes. Here, the energy difference of the host and guest states seemed to be the key factor in order to reduce the impact of field quenching.

2.5.3.3 Summary

In this section, some experimental results on the impact of field quenching in conjugated polymers and small molecule materials has been discussed. For both small molecule and conjugated polymer materials, the photoluminescence can be quenched by up to 80 %, which demonstrates that field quenching is an important loss channel in electrically pumped organic laser diodes. Dye-doped small molecule materials were less affected by field quenching, which makes them interesting for application as active material in organic laser diodes.

2.5.4 Summary

Under intense electrical excitation, various induced loss processes become important in organic semiconductor materials. Therefore, these losses have to be considered when investigating the feasibility of organic laser diodes. Most processes reduce the singlet exciton density, which is the species responsible for optical gain. All loss processes are linked to the low conductivity of organic semiconductors. Due to the low charge carrier mobility, huge particle densities are required in order to achieve current densities above 1 kA/cm². Therefore, bimolecular annihilations between excitons and polarons become important. High polaron and triplet exciton densities also lead to induced absorption processes, which pose additional contributions to the attenuation of the laser mode. In addition, due to the high electric field, excitons can be dissociated to polarons.

2.6 Organic semiconductor lasers

The effect of light amplification by stimulated emission of radiation (LASER) has been first predicted by Einstein based on thermodynamic considerations. The first demonstration of laser operation has been achieved by Maiman with ruby crystals [134]. Optically excited laser action in organic materials, such as dye-doped polymers, dye-doped single crystals and pure anthracene, has already been shown between 1967 and 1974 [135, 136]. However, the fabrication of single crystals is demanding. Since the 1990s, high quality disordered organic semiconductors based on conjugated molecules are available, which can be easily processed from solution or by vacuum evaporation [34]. Organic lasers based on conjugated molecules in solution have been shown in 1992 [137]. In 1996, the first organic semiconductor solid-state laser has been presented [33]. Since then, organic semiconductor lasers (OSL) have been demonstrated for the entire visible spectrum, for various material classes and for different resonator geometries. OSLs have been demonstrated on flexible substrates [138] and low-threshold operation has been demonstrated [139]. An excellent review on the recent progress in the field of OSLs can be found in [14].

In section 2.6.1, the basics of optical gain in conjugated molecules is discussed. Concepts for optically pumped organic semiconductor lasers are presented in section 2.6.2. This section is concluded with a summary in section 2.6.3.

2.6.1 Optical gain in conjugated materials

In figure 2.21, the basic components of a laser are shown schematically. Basically, a laser consists of an optical amplification medium (gain medium) and an optical feedback system (laser resonator or cavity). For laser operation, a population inversion between two states is required. Due to decreasing pumping efficiencies, laser operation is practically not achieved for a two-level medium, while laser operation is possible for three-level and four-level systems. Many gain media exhibit a four-level electronic structure, which allows population inversion even for low pump intensities, hence low-threshold operation can be obtained.

The electronic scheme of a four-level system is sketched in figure 2.22 on the left-hand side. Here, the radiative transitions (1) and (3) are represented by thick arrows. Thin arrows indicate the nonradiative transitions (2) and (4). By optical or electrical excitation (pumping), molecules of the gain medium are excited from the ground state $|1>$ to an excited state $|2>$ from where they quickly relax to a third state $|3>$. In order to ensure high pumping efficiencies, state $|2>$ has to be virtually unpopulated. Therefore, the inverse lifetime of the transition from state $|2>$ to state $|3>$ has to be much higher than the pump rate. The transition (3) is the laser transition. Molecules in state $|3>$ decay radiatively to state $|4>$ by spontaneous recombination. Under high excitation, a population inversion of state $|3>$ and $|4>$ can occur. In this case, stimulated emission is achieved, where a photon initiates a stimulated decay of an excited molecule, which emits a second photon exhibiting the same wavelength, polarization, phase and direction as the first photon. In order to achieve population inversion, the spontaneous lifetime of transition (3) should be high, which ensures a high population density state $|3>$. In order to ensure a high efficiency for stimulated emission, the lifetime of transition (4) should be small in order to depopulate state $|4>$.

The energy scheme for organic laser materials is sketched in figure 2.22 on the right-hand side. In organic semiconducting materials, optical gain is formed by excitons. Thus the laser levels basically correspond to the excitonic energy levels, which have already been discussed in section 2.2. Organic laser materials can be excited optically from the excitonic ground state S_0 to a higher vibronic sideband of the first excited singlet state $|S_1>$, from where it quickly relaxes to the lowest vibronic sideband. The molecule can now undergo stimulated emission to a higher vibronic sideband of the $|S_0>$ band, from where it quickly relaxes to the ground state. Since the relaxation transitions (2) and (4) typically occur on a much shorter timescale than the laser transition (3), organic semiconducting materials are four-level laser systems.

2.6.2 Laser structures

For optical feedback, various resonator geometries have been used for organic semiconductor lasers such as planar microcavities [12, 128, 140–143], Fabry-Perot cavities [34, 144 148], microring [149–151], microdisk [35, 150, 152] and microspherical resonators [153]. A feedback geometry, which is especially interesting for organic semiconductor lasers, are so called distributed feedback (DFB) resonators [16, 154–157]. For DFB resonators, the lasing wavelength can be tuned by adjusting the period within the range of optical gain of the considered material. Since organic laser materials are available for the entire visible spectrum, arbitrary wavelengths are available. Tuning of the laser wavelength can also be done by changing the film thickness in order to adjust the effective refractive index of the waveguide [158, 159] or by mechanical stretching of a flexible sample [160, 161].

Figure 2.23 schematically shows the structure of an organic DFB laser. The slab

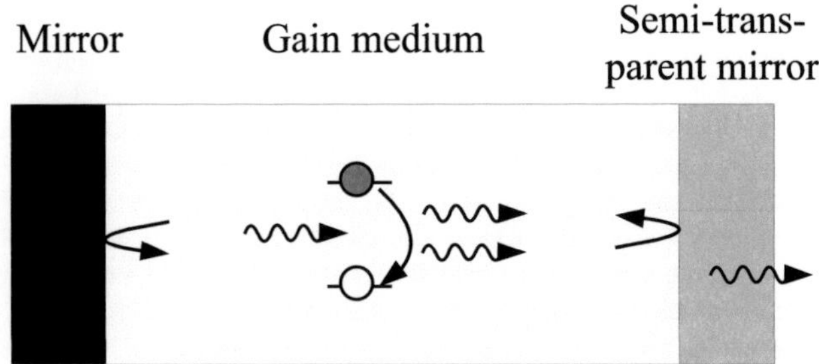

Figure 2.21: Schematic illustration of a laser, consisting of an amplifying material and two mirrors, which form a resonator.

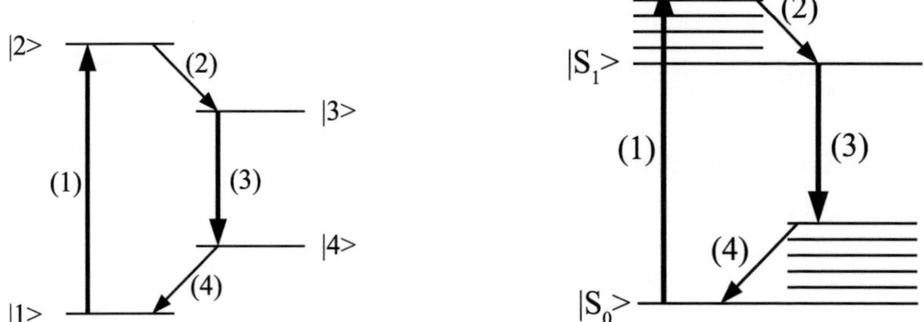

Figure 2.22: Left: Schematic illustration of a four-level laser system. Radiative transitions are indicated by thick arrows. Right: Energy scheme of an organic laser material. For details see text.

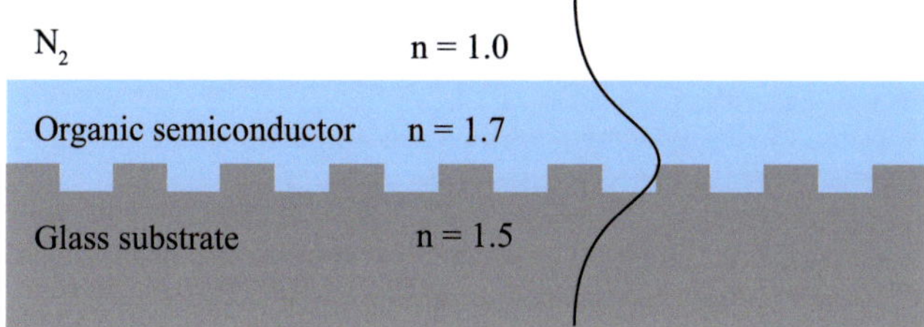

Figure 2.23: Schematic illustration of an organic distributed feedback laser which consists of a structured glass substrate and an organic semiconductor as the active material. Optical feedback is generated by the periodic modulation of the refractive index.

waveguide consists of a glass substrate and a layer of an organic semiconductor material. A periodic structure (grating) is written into the glass substrate by using various lithographic techniques, such as electron lithography, laser interference lithography or direct laser writing in combination with standard photoresist and etching technologies. Then, an organic layer is deposited by vacuum evaporation or solution processes. Since the organic layer exhibits the highest refractive index, a waveguide is formed. In the direction of the waveguide, the index variation generates feedback. For vacuum evaporation, the index variation also exists at the interface between the organic layer and the inert gas (e.g. N$_2$). By altering the length of the grating period, the laser wavelength is defined. For DFB lasers, two laser modes with low threshold exist, where the maxima of the laser modes are either in the glass substrate or in the polymer.

2.6.3 Summary

In this section, the formation of optical gain in organic semiconductors by excitons has been discussed. The concept of distributed feedback resonators is explained, which have been employed successfully for low-threshold organic semiconductor lasers. An overview of feedback structures is given, which have been employed as resonators for organic lasers.

2.7 Concepts for organic laser diodes

Until today, an electrically pumped organic semiconductor laser has not been shown yet. Potential devices for an organic injection laser have to combine a low attenuation laser waveguide with the ability to support high current densities. Highly conductive electrodes exhibit high absorption coefficients. Therefore, the waveguide mode has to be separated from the electrodes. On the other hand, the conductivity and the charge carrier mobilities of organic semiconductors is orders of magnitude lower than in inorganic semiconductors. Hence, huge particle densities are required in order to achieve current densities of the order of 1 kA/cm^2.

In this section, we discuss some published concepts for organic laser diodes, which try to fulfill the abovementioned requirements. First, we discuss double heterostructure organic LEDs as a potential laser diode design in section 2.7.1. Section 2.7.2 analyzes the concept of organic field-effect transistors as a potential device design for organic laser diodes. An interesting device has been proposed by the IMEC group [162], which is analyzed in section 2.7.3. Exciton accumulation may enable the application of low-intensity organic LEDs for use as optical pump sources. This concept is analyzed in section 2.7.4. The section is concluded with a brief summary in section 2.7.5.

2.7.1 Double heterostructure organic LEDs

The main advantage of OLED based design concepts for organic laser diodes is the simple device structure, which enables both easy and reliable fabrication. These devices consist of multiple layers of organic and inorganic materials. These devices are, in principle, one dimensional structures. In addition, it has been shown that these devices can sustain very high current densities of more than 10 kA/cm^2 [25, 64, 163].

In figure 2.24 and in figure 2.25, two potential structures for an organic injection laser are shown, which are based on a three layer double-heterostructure organic LED design. The device consists of transport layers for electrons and holes and an emission layer, where electrons and holes recombine and excitons are formed. In order to allow exciton confinement to the emission layer, a double heterostructure is used. Charge carriers are blocked by the opposite transport layer in order to increase the recombination efficiency. In order to assure high current stability, highly conductive electrodes are used.

For the first design in figure 2.24, the organic double-heterostructure is sandwiched between metal electrodes. A waveguide for the laser mode is formed by the organic layers, which enables the propagation of the TE$_0$-mode. It has been shown by simulation that the quality of this waveguide is of crucial importance for laser operation [164]. A low attenuation coefficient of the waveguide has been found to be inevitable for laser operation, which has also been demonstrated experimentally for optically pumped devices with highly conductive electrodes [165]. The organic layers are assumed to be almost transparent at the laser wavelength (which can be achieved by using materials with a small overlap between the emission and the absorption spectra). The electrodes exhibit high absorption coefficients of the order of 10^6 cm^{-1}, which is a typical value for metals. Hence, the attenuation coefficient of the waveguide at the laser wavelength is determined by the filling factor of the laser mode in the electrode layers. By increasing the thickness of the transport layers from 100 nm to 600 nm, the confinement factor of the laser mode in the electrode layers is strongly decreased [164, 166]. Therefore, the attenuation coefficient is reduced by two orders of magnitude from 1000 cm^{-1} to 10 cm^{-1}. While thick transport layers

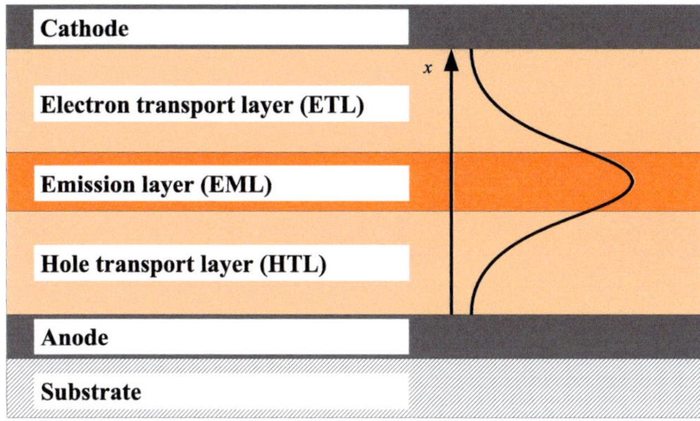

Figure 2.24: Schematic illustration of an organic laser diode structure based on an OLED design: Organic double-heterostructure with thick transport layers for propagation of a TE_0 laser mode.

are advantageous regarding the optical properties, the operation voltage is strongly increased. For 600 nm thick transport layers, voltages of several kV are required in order to achieve current densities of the order of kA/cm^2 [164].

In figure 2.25, an alternative design for an OLED based organic laser diode is presented, which combines low waveguide losses and thin organic layers [167]. The considered organic double heterostructure (DH) device is formed by transport layers for holes (HTL) and electrons (ETL) and an emission layer (EML). The layer thickness is 150 nm for the transport layers and 20 nm for the emission layer. Excitons are assumed to be confined to the emission layer and electrons and holes are blocked by the opposite transport layer, respectively. Transparent conductive oxides are used as electrodes exhibiting attenuation coefficients of 150 cm^{-1} (anode) and 750 cm^{-1} (cathode) [168]. Optical buffer layers are employed enabling the propagation of the TE_2-mode. Since the electrodes are within the minima of the modal intensity profile, waveguide losses of about 4.3 cm^{-1} can be achieved [167].

2.7.2 Double heterostructure organic field-effect transistors

Most difficulties in organic semiconductor materials under high excitation arise from low charge carrier mobilities. In organic field-effect transistors (OFET) however, charge carrier mobilities can be increased dramatically by several orders of magnitude for both n-type [169–172] and p-type [173–175] organic semiconductor materials. For charge carrier mobilities of up to 1 cm^2/Vs, the impact of carrier-related losses would be reduced enormously under high current densities [176–178]. In unipolar OFETs, only one charge carrier is efficiently transported while the opposite charge carrier is restricted to a region very close to the injecting electrode. Hence, excitons are greatly

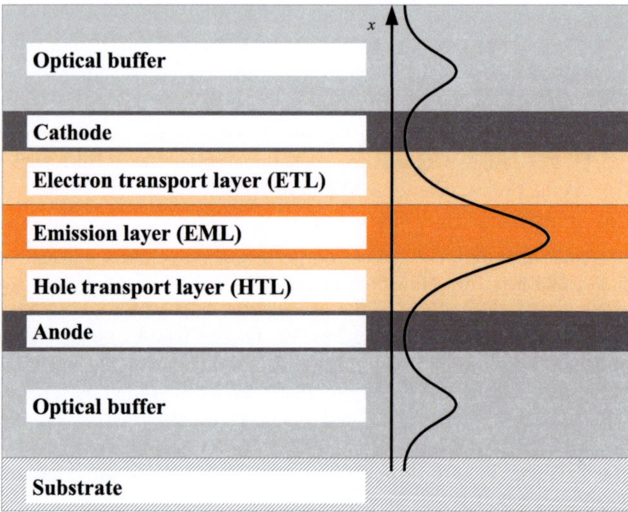

Figure 2.25: An organic laser diode structure based on an OLED design, which is designed for propagation of a TE_2-mode (schematic illustration, adapted from [167]). This design combines thin organic transport layers and a low attenuation coefficients of the waveguide.

quenched by the electrode and some charge carriers of the species exhibiting the higher mobility will not recombine. Thus, the quantum efficiency for light generation is very low.

Recently, transport of both electrons and holes (ambipolar transport) has been demonstrated in OFETs [179–183]. In addition, current densities of up to 100 A/cm^2 have been achieved [184]. Such ambipolar OFET devices show electroluminescence. Under high excitation, optical gain could be generated, since the impact of most charge carrier related loss processes should also be reduced by orders of magnitude. Figure 2.26 sketches the lateral view of an ambipolar (AM) OFET structure. The organic layers are electrically isolated from the gate electrode by a layer of SiO$_2$ or an organic insulator. On top of the insulator, the source and drain electrodes are structured. Then, an organic layer is deposited, which is capable of ambipolar transport (alternative designs with electrodes on top of the organic layer also have been demonstrated). The voltage between the gate and the source electrode is denoted U_{GS}, the voltage between the source and the drain electrode is referred to as U_{SD}. When U_{GS} is applied, an accumulation channel is formed within the organic semiconductor near the organic-insulator interface. When a voltage U_{SD} is applied, both electrons and holes can be injected. The gate voltage U_{GS} can be used in order to tune the conductivity for the charge carriers as well as the position of the recombination zone [185–187]. Hence, the recombination zone can be shifted away from the contact electrodes, which strongly increases the quantum efficiency for light generation. Since the recombination zone can be separated from the source and drain electrodes by several µm, low-loss waveguiding can be achieved.

For OFETs exhibiting a single organic layer, the actual position of the recombination zone also depends on U_{SD}. Under pulsed operation and for high current densities, the position of the recombination zone will be affected. Therefore, the recombination zone should be separated from the electrodes by a proper device design. In order to improve the device performance, double-heterostructure OFETs are discussed. For these devices, the properties of electron and hole transport as well as the emission characteristics can be tuned independently. Therefore, tailored device features can be achieved.

2.7.3 Organic LED with field-enhanced electron transport

Recently, Schols et al. have presented a hybrid device [162, 188, 189], which combines an organic light emitting diode and an organic field effect transistor. This device is sketched in figure 2.27. The device is fabricated on an glass substrate, where an ITO electrode is deposited. Then, an insulating layer is evaporated using a shadow-mask technique. Subsequently, an organic double-heterostructure is deposited, which consists of a hole transport layer, an emission layer and an electron transport layer. The cathode is fabricated on top of the emission layer, whereby the distance from the edge of the emission layer and the cathode is of the order of some µm.

Under electrical excitation, electrons as well as holes are injected into the transport layers due to the applied electric field. Since the anode also functions as a gate

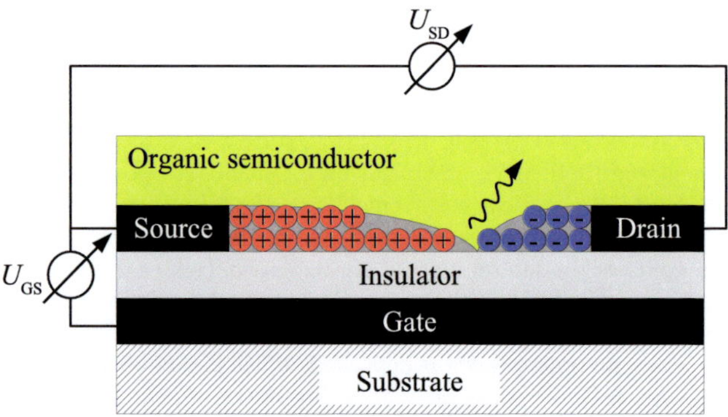

Figure 2.26: Lateral view of an ambipolar organic field-effect transistor (schematic).

electrode, an accumulation channel is formed near the insulator. Hence, the mobility of the electrons is greatly increased and they can be transported parallel to the insulator. Electrons and holes recombine at the edge of the insulator. Since the position of the recombination zone is determined by the device architecture and not by the field mobility of the charge carriers, light emission occurs independently from the applied voltage. The charge carrier and exciton dynamics of the considered device have been studied in detail using a numerical drift-diffusion model [162].

In the considered device, the top electrode is separated from the position of the charge carrier recombination zone. Therefore, both low waveguide attenuation as well as thin organic layers can be combined simultaneously in this structure, since the laser mode can be separated from the electrodes. Hence, the presented structure is of interest as a potential design for an organic semiconductor laser diode.

2.7.4 Exciton accumulation

For organic laser diode design concepts based on OLED or OFET structures, luminescence is achieved by direct electrical pumping, where excitons are generated by Langevin recombination of electrons and holes. However, the simultaneous presence of triplet excitons and polarons in the emission layer cause various quenching and absorption processes to become important. Hence, optical pumping using compact pumping sources are an alternative in order to provide compact tunable organic semiconductor laser (OSL) sources. Recently, low-threshold organic laser devices have been demonstrated which have enabled optical pumping using inorganic laser diodes [19–22]. The attempt to pump an OSL using an OLED was not successful [24].

A different approach has been presented by Giebink et al., who studied the enhanced photoluminescence by accumulation of field-stabilized geminate charge pairs [190,191]. For an accumulation factor of 1000, singlet exciton densities of the order of

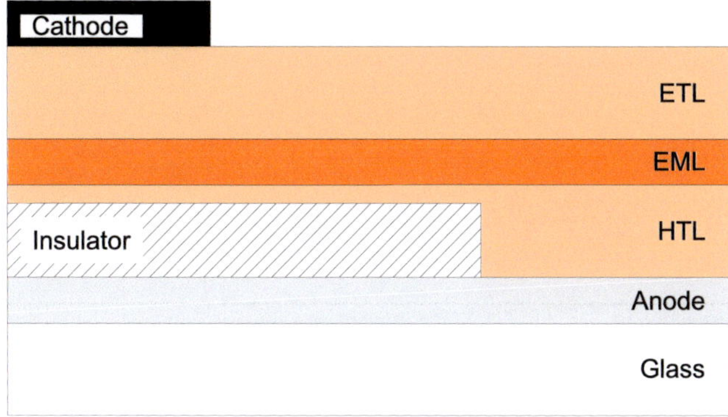

Figure 2.27: Lateral view of an organic light emitting diode with field-enhanced electron transport (adapted from [162]).

10^{17} /cm^3 would be achieved, which is the lower limit for the singlet exciton density in organic lasers.

In figure 2.28, the considered device is sketched, which consists of a 100 nm thick emission layer. In the emission layer, a small molecule guest-host system is employed, where the laser dye coumarin 6 (C6) is doped into 4,4-bisN-carbazolylbiphenyl (CBP). The CBP:C6 layer is sandwiched between two layers of Teflon AF^{TM}. An ITO anode and an Aluminum cathode are employed.

A voltage of 40 V is applied between the ITO and the aluminum electrodes. Then, the organic emission layer is pumped at power density of 1 W/cm^2, which is of the same order of magnitude as the power density of high-brightness OLEDs [192]. Due to the electric field, some of the excitons are dissociated and geminate electron hole pairs are generated. In figure 2.29, the energy levels of the ground state S_0 and the first excited singlet state S_1 of the CBP and the C6 are shown. In this case, the hole is trapped at the C6, since the energy of the ground state is higher than the ground state of the CBP. However, the electron is mobile within the attractive potential of the hole.

When the voltage is removed, the geminate charge pairs recombine and a burst of the luminescence intensity is achieved. The magnitude of the burst increases for increasing pumping time periods. An increase in photoluminescence by a factor of 22 has been demonstrated for $\tau = 8$ µs. For longer pumping, further increase is not achieved due to nongeminate recombination and a decrease of the internal electric field due to charge buildup from fully dissociated charge pairs.

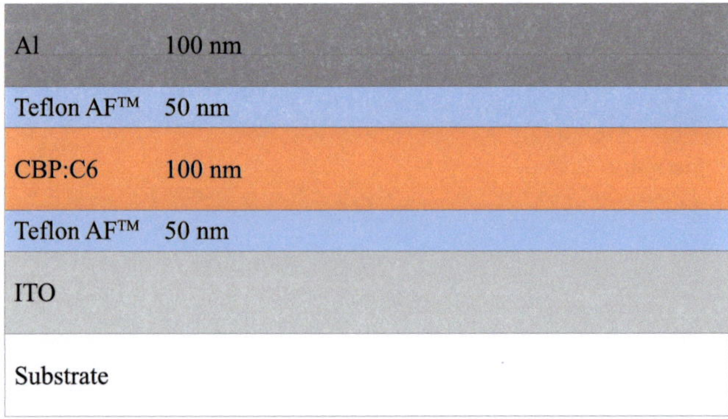

Figure 2.28: Geminate polaron pair accumulation device structure consisting of a coumarin 6 (C6) doped 4,4-bisN-carbazolylbiphenyl (CBP) emission layer, which is sandwiched between two Teflon AF^{TM} layers (adapted from [190, 191]).

Figure 2.29: Energy band scheme for the guest-host system of C6 doped CBP (adapted from [190, 191]). The hole is trapped on the C6 while the electron is mobile within the attractive potential of the hole.

2.7.5 Summary

In this section, several concepts for an organic laser diode have been discussed. While device designs, which are based on the OLED concept, allow very high current densities of several kA/cm^2, various loss processes become important under high excitation due to low charge carrier mobilities. Higher charge carrier mobilities can be achieved in ambipolar, bilayer or double-heterostructure OFET devices. In OFET structures, the emission layer is separated from the electrodes. Hence, lower waveguide attenuation might be achieved. Hybrid device designs have been discussed, which combine advantages of OLEDs and OFETs. In a further approach, geminate polaron pair accumulation by field stabilization has been demonstrated, which would allow pumping of an OSL using a low-intensity light source such as an OLED. However, for none of the discussed device structures electrically pumped laser operation has been shown yet.

2.8 Chapter summary

In this chapter, the basics of organic semiconductor materials has been discussed and devices based on organic semiconductors, such as organic light emitting diodes and organic field effect transistors, have been introduced. The advantages and unique features of organic semiconductors have been reconsidered, which enable a number of unique applications and devices. A particularly interesting but so far unrealized device is an organic laser diode. Electrically pumped organic semiconductor laser devices are expected to provide numerous advantages like the tunability over the whole visible spectrum and the promise of low-cost, large area fabrication. Although high current densities have been demonstrated for thin-film devices, neither an electrically pumped organic laser nor an electrically pumped organic amplifier has been shown yet. In addition, promising design concepts for organic laser diodes have been presented. In the following, we focus on the double-heterostructure OLED based design concept, which has been presented in section 2.7.1.

3 Modelling of organic laser diodes

In this work, organic laser diode structures are considered, which consist of a multi-layer organic double-heterostructure. The device performance is influenced by both electrical properties, which determine the dynamics of charge carriers and excited states in the device, and optical properties, which influence the quality of the laser waveguide. Thus, the characteristics of the device is influenced by more than 50 parameters. Hence, a purely experimental optimization of the device is cumbersome, time consuming and basically impossible.

In order to develop design rules and in order to support the experimental work, numerical simulations are used. For these simulations, the electrical and optical features of the organic laser diode structure have to be modeled. In this chapter, the employed model is presented, which describes the electrical properties of the device by a one-dimensional self-consistent drift-diffusion model and which also accounts for the following processes: field-dependent injection of charge carriers, drift and diffusion of particles, Langevin recombination, annihilation processes between particles, singlet and triplet excitons, polaron absorption, triplet absorption, exciton dissociation, stimulated emission and recombination. In addition, the optical properties of the waveguide are modeled using a transfer matrix method.

This chapter is structured as follows: In section 3.1, the model for the device geometry is presented. In section 3.2 and section 3.3, the models describing the optical and electrical properties are discussed. Loss processes at high excitation are considered in section 3.4. The chapter is summarized in section 3.5.

3.1 Device geometry

For the realization of an organic laser diode, several device architectures are under consideration, which have been discussed in section 2.7. One promising structure is a double-heterostructure (DH) organic light-emitting diode (OLED), which is being considered as a candidate for an actual laser diode. This work focuses on the fundamental mechanisms in DH-OLEDs. Different device designs such as organic field-effect geometries are not analyzed in this work. In this section, the modelling of the device geometry is discussed.

Figure 3.1 shows the one dimensional geometrical model of a laser diode employing an OLED geometry. The device exhibits an organic double-heterojunction, which consists of a hole transport layer (HTL), an emission layer (EML) and an electron transport layer (ETL). The organic layers are sandwiched between two conductive electrodes (anode for hole injection, cathode for electron injection). As conductive

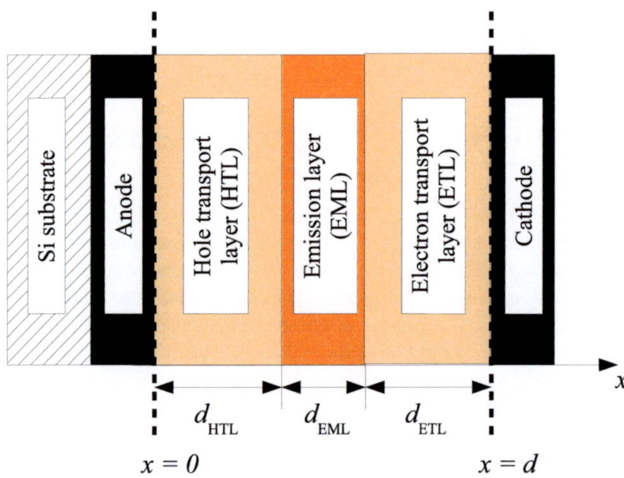

Figure 3.1: Device geometry for a double-heterostructure organic laser diode, consisting of an electron transport layer (ETL), emission layer (EML) and a hole transport layer (HTL). Each layer is specified by a one dimensional grid. At each grid point, the time dependent local field, particle densities, currents and reaction rates are calculated.

materials, metals or transparent conductive oxides are employed. Depending on the actual device design, additional layers are employed beyond the electrodes.

The thickness of the organic layers is of the order of 300 nm, while the lateral dimensions are of the order of some 100 µm. Thus, the active area of the device can be described by a one-dimensional model. This simplification greatly reduces the computational complexity and enables the simulation of an entire device on a workstation. Each layer of the device is specified by several parameters, which describe its optical and electrical properties and account for the impact of various loss processes.

3.2 Modelling of the optical properties

Organic laser diode structures have to exhibit a waveguide for the laser mode. The actual modal intensity profile of the guided laser mode has to be considered in order to determine the attenuation coefficient of the waveguide as well as the rate for stimulated emission. The modal intensity profile is calculated by using a transfer-matrix method, which is presented in this section.

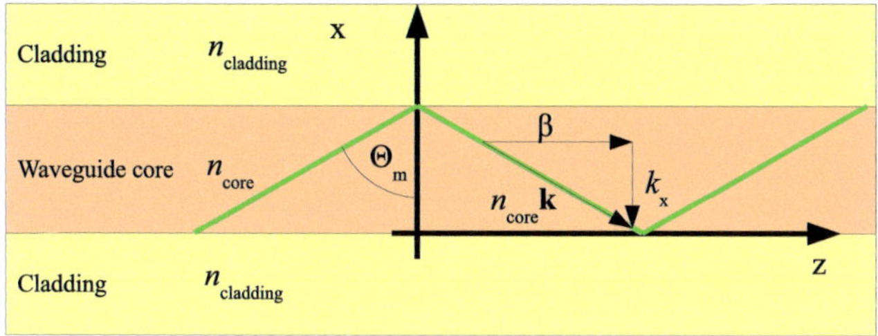

Figure 3.2: Sketch of waveguiding in a slab waveguide structure consisting of a high-index core (n_{core}) and a low-index cladding (n_{cladding}). Waveguiding is achieved by total reflection, if the angle of incidence Θ_{m} is larger than the critical angle for total reflection Θ_{c}.

3.2.1 Wave guiding

For optical feedback and amplification, a waveguide for the laser mode is required. Waveguiding is achieved for slab devices, where a high-index layer is surrounded by lower index materials. Regarding organic laser diodes based on an OLED design, the modal intensity profile has to be calculated in order to determine the interaction between photons of the laser mode and charge carriers as well as excited states. Interaction mechanisms are stimulated emission, where optical gain is generated, and particle induced absorption processes, such as polaron and triplet-triplet absorption.

When the thickness of the waveguide is much larger than the optical wavelength, the propagation of light can be described by classical optics. For organic thin-film structures, however, the total thickness of the waveguide is less than 1 µm. Thus, wave optics has to be employed. Waveguiding is achieved by total reflection at the interface of a high-index waveguide core (refractive index is n_{core}) and a low-index waveguide cladding (refractive index is n_{cladding}). Waveguiding is achieved, if the angle of incidence Θ_{m} is larger than the critical angle for total reflection Θ_{c}:

$$\sin \Theta_{\text{m}} \geq \frac{n_{\text{cladding}}}{n_{\text{core}}} = \sin \Theta_{\text{c}}. \tag{3.1}$$

Guided modes can be identified by employing the transverse resonance condition:

$$2 k_x n_k d_{\text{core}} \cos \Theta_m - 4\phi_{\text{cladding}} = m \times 2\pi, \tag{3.2}$$

where d_{core} is the thickness of the core, k_x is the component of the wave vector **k** in x direction and Φ_{cladding} is the phase shift for total reflection at the interfaces of waveguide core and waveguide cladding. m is a positive integer and the index of the mode. Equation (3.2) can only be solved for an finite number of values of m. The propagation constant β is the component of the wave vector **k** in z direction.

Waveguiding within an organic laser diode is sketched in figure 3.3. Here, a high-

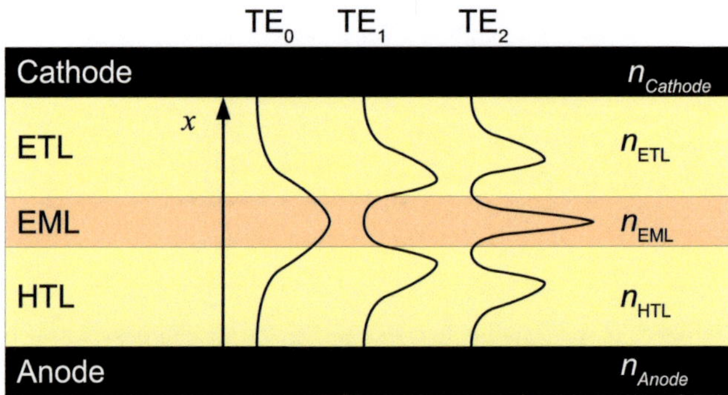

Figure 3.3: Schematic lateral view of a waveguide within an organic laser diode. For low-attenuation waveguiding, the refractive index of the emission layer has to exceed the refractive index the transport layers.

index emission layer is surrounded by lower index transport layers. For isotropic and linear materials, the propagation of electromagnetic waves is described by the one-dimensional Helmholz equation

$$\nabla^2 F\left(x\right) - k_0^2 \sqrt{n} F\left(x\right) = 0, \tag{3.3}$$

where $k_0 = \frac{2\pi}{\lambda}$ is the wave number and n is the refractive index. The solutions of the Helmholtz equation have to decrease exponentially for $x \to \pm\infty$ and have to be continuous (boundary conditions). In addition, the polarization of the wave has to be considered. For planar wave guides, guided TE- (the electric field F and the plane of incidence are perpendicular) and TM- modes (the electric field F and the plane of incidence are coplanar) have to be considered. For guided modes, the intensity profile have to decrease exponentially for $x \to \pm\infty$ (boundary conditions). Depending on the thickness of the waveguide and its refractive index, multiple modes can propagate. In figure 3.3, the modal intensity profile of the modes TE$_0$, TE$_1$ and TE$_2$ are sketched.

3.2.2 Transfer-matrix method

In our model, the Helmholtz equation is solved using a transfer-matrix method. The transfer-matrix method has been employed successfully for describing the optical properties of one-dimensional devices [193]. This method can be employed for slab thin-film structures of isotropic and linear homogeneous materials.

Figure 3.4 sketches the employed transfer matrix method for the calculation of the modal intensity profile in x direction. The considered device is discretized into cells of equal thickness d. The optical characteristics of each layer is described by the

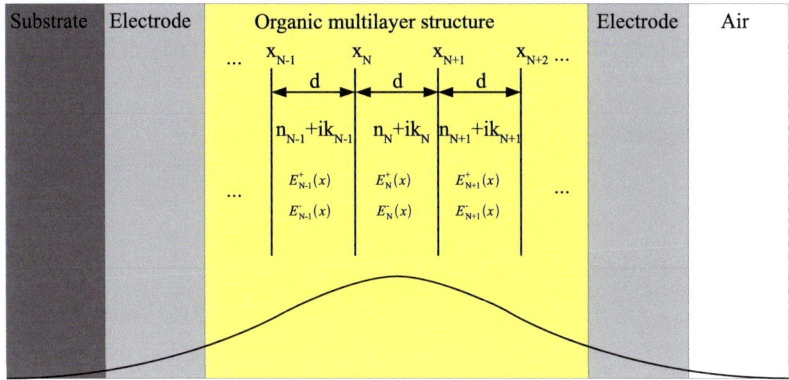

Figure 3.4: A transfer matrix method is employed in order to calculate the modal intensity profile of the laser modes.

complex refractive index $\tilde{n} = n + ik$. Here, n denotes the real part of the refractive index and k its imaginary part. n and k are functions of the laser wavelength. The imaginary part of the refractive index k is related to the material attenuation coefficient of the medium α by the following relation

$$\alpha = \frac{4\pi k}{\lambda}. \tag{3.4}$$

At the interfaces between two organic layers, photons can either be reflected or transmitted. Therefore, a separation of the electric field into two components E^+ and E^- is practicable, which propagate in positive and negative x direction respectively. At the interface of two cells N and $N + 1$ at $x = x_{N+1}$, the probability for reflection $r_{N,N+1}(x_{N+1})$ and transmission $t_{N,N+1}(x_{N+1})$ is determined by

$$r_{N,N+1}(x_{N+1}) = \frac{E_N^-(x_{N+1})}{E_N^+(x_{N+1})} = \frac{n_N - n_{N+1}}{n_N + n_{N+1}}, \tag{3.5}$$

$$t_{N,N+1}(x_{N+1}) = \frac{E_{N+1}^+(x_{N+1})}{E_N^+(x_{N+1})} = \frac{2 \times n_{N+1}}{n_{N+1} + n_N}. \tag{3.6}$$

Here, $E_N^+(x_{N+1})$ and $E_N^-(x_{N+1})$ denote the complex electric field component at the position x within cell N.

From equations (3.5) and (3.6), a boundary condition can be derived for the electric field at the interface of two cells:

$$E_N^+(x_{N+1}) = \frac{1}{t_{N,N+1}} E_{N+1}^+(x_{N+1}) + \frac{r_{N,N+1}}{t_{N,N+1}} E_{N+1}^-(x_{N+1}) \tag{3.7}$$

By applying the equation in reverse direction, the reverse relation is derived. For better readability, the matrix notation is employed. At the interface, the electric field in both cells can be calculated using

$$\begin{bmatrix} E_N^+ (x_{N+1}) \\ E_N^- (x_{N+1}) \end{bmatrix} = I_{.N,N+1} \times \begin{bmatrix} E_{N+1}^+ (x_{N+1}) \\ E_{N+1}^- (x_{N+1}) \end{bmatrix},$$ (3.8)

where the interface matrix is

$$I_{.N,N+1} = \begin{bmatrix} \frac{1}{t_{N+1,N}} & \frac{r_{N+1,N}}{t_{N+1,N}} \\ \frac{r_{N+1,N}}{t_{N+1,N}} & \frac{1}{t_{N+1,N}} \end{bmatrix}.$$ (3.9)

The propagation of the electromagnetic wave within a cell is described by the propagation matrix

$$P_N (x) = \begin{bmatrix} \exp\left(-ik_x \widetilde{n}_N x\right) & 0 \\ 0 & \exp\left(ik_x \widetilde{n}_N x\right) \end{bmatrix},$$ (3.10)

where k_x is the component of the wave vector in x direction (see figure 3.2) and $\widetilde{n}_N$ is the complex refractive index of layer N.

3.2.3 Photon density

The areal photon density q is affected by stimulated emission, waveguide absorption, induced polaron absorption, induced triplet exciton absorption and by spontaneously emitted photons coupling into the laser mode. The rate equation for q is

$$\begin{aligned}
\frac{\partial q(t)}{\partial t} &= -\frac{q(t)}{\tau_{cav}} + \frac{\beta}{\tau_{rad}} \int_0^d n_{S_1}(x,t)\, dx \\
&+ k_{SE} q(t) \int_0^d \hat{I}(x)\, n_{S_1}(x,t)\, dx \\
&- k_{carrier} q(t) \int_0^d \hat{I}(x) \left(n_e(x,t) + n_h(x,t)\right) dx \\
&- k_{T_1 T_N} q(t) \int_0^d \hat{I}(x)\, n_{T_1}(x,t)\, dx.
\end{aligned}$$ (3.11)

Here, the first term on the right hand side describes the waveguide losses. The second term describes the coupling of spontaneously emitted photons into the laser mode. Stimulated emission is expressed by the third term. The last two terms stand for the impact of induced absorptions by charge carriers and triplet excitons, respectively.

3.3 Electrical modelling

In this section, the electrical model is discussed. The electrical properties of the organic device are described within the drift-diffusion approximation. When a voltage is applied to the electrodes of the device, charge carriers are injected into the organic layers. The charge carriers are transported by drift and diffusion and eventually undergo Langevin recombination. Singlet and triplet excitons are formed, which can undergo radiative or nonradiative decay.

3.3.1 Rate equations for particles

The dynamics of charge carriers and excitons are described by a one-dimensional self-consistent drift-diffusion model which also accounts for the following processes: field-dependent injection of charge carriers, drift and diffusion of particles, Langevin recombination, annihilation processes between charge carriers, singlet and triplet excitons, spontaneous and stimulated emission, recombination, exciton dissociation, polaron absorption and triplet absorption. The model consists of continuity equations for the hole (n_h), electron (n_e), singlet exciton (n_{S_1}) and triplet exciton (n_{T_1}) densities inside the organic semiconductor layers:

$$\frac{\partial n_{\mathrm{e,h,S_1,T_1}}(x,t)}{\partial t} + \frac{\partial \Gamma_{\mathrm{e,h,S_1,T_1}}(x,t)}{\partial x} = S_{\mathrm{e,h,S_1,T_1}}(x,t) \tag{3.12}$$

S is the source term that describes the generation and recombination of particles. The formation of excitons from holes and electrons is assumed to be a Langevin process. In addition, singlet excitons can undergo stimulated emission and excitons can be deactivated by bimolecular annihilation processes. Under high fields, excitons can be dissociated to free charge carriers. Thus, the source term S contains the following contributions

$$S_{\mathrm{e,h,S_1,T_1}}(x,t) = \left.\frac{\partial n_{\mathrm{e,h,S_1,T_1}}(x,t)}{\partial t}\right|_{\mathrm{LR}}$$

$$+ \left.\frac{\partial n_{\mathrm{S_1,T_1}}(x,t)}{\partial t}\right|_{\mathrm{decay}}$$

$$+ \left.\frac{\partial n_{\mathrm{S_1}}(x,t)}{\partial t}\right|_{\mathrm{SE}} \tag{3.13}$$

$$+ \left.\frac{\partial n_{\mathrm{S_1,T_1}}(x,t)}{\partial t}\right|_{\mathrm{BA}}$$

$$+ \left.\frac{\partial n_{\mathrm{e,h,S_1,T_1}}(x,t)}{\partial t}\right|_{\mathrm{FQ}}. \tag{3.14}$$

The terms on the right hand side describe the impact of Langevin recombination (LR), radiative and nonradiative monoexponential decay (decay), stimulated emission (SE), bimolecular annihilation processes (BA) and field-induced exciton dissociation (field quenching, FQ). The impact of Langevin recombination, radiative and nonradiative monoexponential decay and stimulated emission on the particle densities are discussed in the following sections. The impact of the loss processes bimolecular annihilation and field quenching on the particle densities are discussed in a more elaborate manner in section 3.4.1 and section 3.4.2.

3.3.2 Drift-diffusion model

The particle fluxes Γ_{e,h,S_1,T_1} are given by the drift-diffusion approximation

$$
\begin{aligned}
\Gamma_{e,h,S_1,T_1}(x,t) &= n_{e,h}(x,t)\mu_{e,h}(F)F(x,t) \\
&\quad - D_{e,h,S_1,T_1}(F)\frac{\partial n_{e,h,S_1,T_1}(x,t)}{\partial x}
\end{aligned}
\tag{3.15}
$$

with Poole-Frenkel type charge carrier mobilities $\mu(F)$. The diffusion coefficients $D_{e,h}(F)$ are calculated from the mobilities using the Einstein relation:

$$
D_{e,h}(F) = \frac{\mu_{e,h}(F)k_B T}{q}.
\tag{3.16}
$$

3.3.3 Charge carrier injection

For modelling of charge carrier injection, the model developed by Scott and Malliaras is employed [54], which has been discussed in detail in section 2.3.4. For the current densities of the charge carriers directed away from the contact, the following expression is employed which has been derived by Scott and Malliaras [54]

$$
J = 4\Psi^2 N_0 e\mu F \exp\cdot\left(-\frac{\Phi_B}{k_B T}\right)\exp f^{0.5},
\tag{3.17}
$$

where the reduced field f is defined

$$
f = \frac{eF r_C}{k_B T},
\tag{3.18}
$$

and

$$
\Psi(f) = f^{-1} + f^{-0.5} - f^{-1}\left(1 + 2f^{0.5}\right)^{0.5}.
\tag{3.19}
$$

Thermionic injection as well as carrier back-flow into the contact due to the low mobilities is included.

3.3.4 Exciton generation

Excitons are generated by Langevin recombination, which has been discussed in section 2.3.3. The impact of Langevin recombination on the particle densities is

$$
\begin{aligned}
\left.\frac{\partial n_{e,h}\left(x,t\right)}{\partial t}\right|_{LR} &= -\frac{e\left(|\mu_e|+\mu_h\right)}{\epsilon_r\epsilon_0}n_e\left(x,t\right)n_h\left(x,t\right) \\
\left.\frac{\partial n_{S_1}\left(x,t\right)}{\partial t}\right|_{LR} &= \xi\cdot\frac{e\left(|\mu_e|+\mu_h\right)}{\epsilon_r\epsilon_0}n_e\left(x,t\right)n_h\left(x,t\right) \\
\left.\frac{\partial n_{T_1}\left(x,t\right)}{\partial t}\right|_{LR} &= \left(1-\xi\right)\cdot\frac{e\left(|\mu_e|+\mu_h\right)}{\epsilon_r\epsilon_0}n_e\left(x,t\right)n_h\left(x,t\right).
\end{aligned}
\tag{3.20}
$$

Here, ξ denotes the probability for singlet exciton generation. In our model, we assume that every fourth generated exciton is a singlet exciton, thus $\xi = 0.25$.

3.3.5 Radiative and nonradiative decay

Singlet and triplet excitons can be deactivated by radiative and nonradiative decay. The total exponential lifetime is

$$
\frac{1}{\tau_{S_1,T_1}} = \frac{1}{\tau_{S_1,T_1,\text{rad}}} + \frac{1}{\tau_{S_1,T_1,\text{nonrad}}},
\tag{3.21}
$$

where the radiative lifetime is $\tau_{S_1,T_1,\text{rad}}$ and the nonradiative lifetime is $\tau_{S_1,T_1,\text{nonrad}}$. The impact of radiative and nonradiative decay on the singlet exciton density is

$$
\left.\frac{\partial n_{S_1,T_1}\left(x,t\right)}{\partial t}\right|_{\text{decay}} = -\frac{n_{S_1,T_1}}{\tau_{S_1,T_1}}.
$$

3.3.6 Stimulated emission

The effect of stimulated emission on the singlet exciton density is

$$
\begin{aligned}
\left.\frac{\partial n_{S_1}\left(x,t\right)}{\partial t}\right|_{\text{stim}} &= -\, k_{\text{stim}}q\left(t\right)I\left(x\right)n_{S_1}\left(x,t\right) \\
k_{\text{stim}} &= -c\cdot\sigma_{\text{SE}}/n_{\text{eff}}.
\end{aligned}
\tag{3.22}
$$

Here, σ_{SE} is the cross section for stimulated emission, $q\left(t\right)$ is the areal photon density and n_{eff} is the effective refractive index of the laser mode.

	Si	Al$_2$O$_3$	HT-AZO	α-NPD	Alq$_3$:DCM	BPhen	AZO	Alq$_3$	Air
d /nm		334	40	150	20	150	40	290	
n	1.46	1.66	1.90	1.81	1.73	1.75	1.90	1.73	1.00
α /cm	0	5	150	0	0	0	750	5	0
$\mu_e/(\mathrm{cm}^2/\mathrm{Vs})$				-	$5 \cdot 10^{-6}$	$5 \cdot 10^{-4}$			
$\mu_h/(\mathrm{cm}^2/\mathrm{Vs})$				$5 \cdot 10^{-4}$	$5 \cdot 10^{-6}$	-			

Table 3.1: Overview of refractive indices, absorption coefficients and charge carrier mobilities

3.3.7 Laser threshold current density

The determination of the threshold current density has been described in detail in [164]. Here, the areal photon density $q(t)$ as a function of the current density j is analyzed. The laser threshold is determined from the laser characteristics. At the threshold current density, $\partial^2 q(j) / \partial^2 j$ exhibits a maximum.

3.3.8 Material parameters

In table 3.1 and table 3.2, the electronic and optical material properties are summarized. Below, the choice of the parameters is motivated.

For organic laser diodes, a multilayer design is necessary in order to achieve sufficient gain for current densities of the order of kA/cm^2. Double heterostructures can be fabricated easily from small molecule materials using vacuum deposition. The fabrication of multilayer devices from polymers is much more challenging. The solvents of the subsequent layers may not dissolve the underlying layers. Hence cross-linkable polymers or materials with orthogonal solvents have to be used and not all polymers can be combined. In this thesis, typical properties of small molecule materials have been selected as a starting point for the parameter study.

The charge carrier mobilities of transport layers are chosen as 5×10^{-4} cm^2/(Vs). Similar mobilities are found for TPD or α-NPD as hole transport materials [194] and for BPhen as electron transport material [195]. Polymers have even higher mobilities of up to 10^{-2} cm^2/(Vs) [157, 196, 197]. In order to diminish exciton quenching by annihilation processes, the singlet excitons should be generated homogeneously all over the active layer. Hence, the mobilities of electrons and holes have to be of the same order of magnitude. For small molecules, doping with dyes is often essential in order to achieve good electroluminescence efficiencies. However, doping often greatly reduces the charge mobilities [198]. Thus, we have assumed that in the emission layer the mobilities of electrons and holes are 100 times smaller than the mobilities in the transport layer. In contrast, polymers can combine good electroluminescence efficiency and both high electron and hole mobilities of up to 10^{-2} cm^2/(Vs) [196,197].

Charge carrier mobilities in many organic semiconducting materials show a good agreement with a Poole-Frenkel law. However, the field dependence of the charge carrier mobility is usually weak in materials with a high mobility. Therefore, the

Electronic properties		Optical properties	
Parameter	Value	Parameter	Value
$\mu_{HTL} = \mu_{ETL}$	5×10^{-4} cm^2/(Vs)	λ	620 nm
$\mu_{EML;h} = \mu_{EML;e}$	5×10^{-6} cm^2/(Vs)	σ_{stim}	7×10^{-16} cm^2
τ_{S_1}	2 ns	β	10^{-5}
τ_{T_1}	25 µs	n_{Si}	3.906-0.022i
$D_{S_1} = D_{T_1}$	5×10^{-4} cm^2/s	n_{Au}	0.13-3.16i
T	300 K	n_{Al}	1.29-7.48i
$\epsilon_{HTL} = \epsilon_{EML} = \epsilon_{ETL}$	4.0	$n_{HTL} = n_{ETL}$	1.7
E_{S_1}	0.4 eV	n_{EML}	1.75
E_{T_1}	0.6 eV	n_{Air}	1

Table 3.2: Overview of the electronic and optical material parameters, which have been employed for our simulations.

field dependence of the mobility has been set to zero in our calculations. Excellent waveguiding and a low attenuation coefficient are mandatory prerequisites for lasing [164]. Hence the refractive index of the organic layers is of great importance. Organic semiconducting materials usually exhibit a refractive index between 1.6 and 1.8 [199, 200]. For polymers as well as small molecule materials, the radiative lifetimes of singlet excitons is in the range of 0.2 ns to 20 ns [89,201]. In contrast, triplet excitons normally exhibit a lifetime in the range from 1 µs to 10 ms. In our simulations, we assume a triplet exciton lifetime of 25 µs [202].

The main parameters for field quenching are the exciton binding energies of the singlet and triplet excitons. Typically, singlet exciton binding energies are in the range from 0.2 eV to 1.4 eV [109–112]. The triplet exciton binding energies are 0.2 eV up to 1.0 eV higher than the singlet exciton binding energies [109,111]. In our simulations, a singlet exciton binding energy of 0.4 eV and a triplet exciton binding energy of 0.6 eV is employed.

The organic double heterostructure is enclosed by electrical contacts. For organic laser diodes, the electrode materials have to exhibit high conductivity in order to enable operation under high current densities. In addition, the attenuation coefficients of the electrode materials at the laser wavelength have to be low in order to reduce the threshold of the laser mode. For the investigated organic laser diode structures, either metals or transparent conductive oxides (TCO) are employed. While metals provide high conductivity, they exhibit attenuation coefficients of about 10^5 /cm. TCOs such as indium tin oxide (ITO) or aluminum doped zinc oxide (AZO) exhibit lower conductivity, but also lower attenuation coefficients. For AZO, which has been processed under high temperatures (HT-AZO), attenuation coefficients as low as 150 cm^{-1} have been demonstrated [168].

3.4 Modelling of loss processes under high excitation

In this section, the modelling of loss processes is described, which occur at high excitation and whose impact is in particular striking for organic laser diodes. Due to the low charge carrier mobility of organic semiconductors, huge charge carrier densities are required in order to achieve current densities of the order of several kA/cm^2. Thus, bimolecular annihilation processes become important. The modelling of their impact is discussed in section 3.4.1. Due to the low conductivity of organic semiconductors, electric fields of several MV/cm are required to drive high current densities. The modelling of field effects on the exciton density is discussed in section 3.4.2. The modelling of induced absorption processes such as polaron and triplet-triplet absorption is discussed in section 3.4.3. The section is concluded with a summary in section 3.4.4.

3.4.1 Bimolecular annihilations

In this section, various annihilation processes between the different particle species are modeled. The following processes are considered: Singlet-singlet (SSA), singlet-triplet (STA), triplet-triplet (TTA), singlet-polaron (SPA) and triplet-polaron annihilation (TPA) as well as intersystem crossing (ISC). The basics of bimolecular annihilation processes have been discussed in section 2.5.1. In this section, rate coefficients for annihilation processes in different materials are presented and the employed parameters are motivated. Following this, the singlet and triplet exciton rate terms for bimolecular annihilation processes are derived.

3.4.1.1 Rate coefficients

Table 3.3 summarizes some published rate coefficients for different polymer and small molecule materials. The rate coefficients for some of the annihilation processes vary by up to three orders of magnitude. As can be seen, none of the materials is completely characterized in terms of annihilation processes.

The lack of measured rate coefficients makes it impossible to model the exact properties of a specific material. For this reason, typical rate coefficients have been selected for use as a starting point in our simulations. The rate coefficients of this starting point can be found in the last row of Table 3.3. It has typical material properties and typical annihilation rate coefficients of actual small molecule materials. The employed rate coefficients are in the range of the smallest constants that can be found in the literature. At this point any dependence of the annihilation parameters on other parameters, such as the mobility of electrons and holes or the diffusion constant of excitons, is ignored.

Material	$\kappa_{SS}/(\mathrm{cm}^3/\mathrm{s})$		$\kappa_{ST}/(\mathrm{cm}^3/\mathrm{s})$		$\kappa_{TT}/(\mathrm{cm}^3/\mathrm{s})$		$\kappa_{SP}/(\mathrm{cm}^3/\mathrm{s})$		$\kappa_{TP}/(\mathrm{cm}^3/\mathrm{s})$		$\kappa_{ISC}/(1/\mathrm{s})$	
Alq$_3$	3.5×10^{-11}	[63, 64]										
DCM2:CBP							3.0×10^{-10}	[203]	1.0×10^{-12}	[203, 204]		
F8BT	2.6×10^{-9}	[205]									6.7×10^6	[206]
LOPP7							9.0×10^{-8}	[115]				
m-LPPP			5.7×10^{-10}	[207]			9.0×10^{-8}	[115]				
LPPP	4.2×10^{-9}	[118, 208]										
MEH-PPV			1.5×10^{-9}	[207]	1.0×10^{-14}	[209]						
PFO	4.9×10^{-8}	[205]	5.2×10^{-10}	[207]								
PIFTEH	8.0×10^{-8}	[83]										
PIFTO	2.5×10^{-8}	[83]										
PPDP			1.6×10^{-9}	[207]								
PPV	6.0×10^{-9}	[65]	1.2×10^{-9}	[207]			1.2×10^{-8}	[65]				
DOO-PPV											5×10^7	[210]
6% PtOEP:Alq$_3$					1.2×10^{-13}	[203, 211]						
8% PtOEP:CBP					3.0×10^{-14}	[203, 212]			1.0×10^{-12}	[211]		
PtOEP:PSF-TAD									4×10^{-13}	[66]		
TPD					1.0×10^{-11}	[107]						
Standard material	10^{-10}		10^{-10}		10^{-12}		10^{-10}		10^{-12}		5×10^7	

Table 3.3: Overview of rate coefficients for various annihilation processes. Values are given in $(\mathrm{cm}^3/\mathrm{s})$, values for κ_{ISC} are given in $(1/\mathrm{s})$.

3.4.1.2 Singlet exciton rate equation for BA

The densities of singlet and triplet excitons are influenced by bimolecular annihilations. In equation (3.23), the singlet annihilation rate term is analyzed:

$$
\left.\frac{\partial n_{S_1}(x,t)}{\partial t}\right|_{BA} = -(2-\xi)\cdot\kappa_{SSA}n_{S_1}^2 - \kappa_{STA}n_{S_1}n_{T_1}
$$
$$
+\xi\kappa_{TTA}n_{T_1}^2 - \kappa_{SPA}(n_e + n_h)n_{S_1}
$$
$$
-\kappa_{ISC}n_{S_1} \tag{3.23}
$$

The rate terms depend on the spatiotemporal densities of the involved particle species and the rate coefficients. The rate terms can be derived from the reaction equations. The first term on the right hand side in equation (3.23) describes the loss rate of singlet excitons due to SSA. Since SSA is a bimolecular process, the rate term is proportional to the square of the singlet exciton density. In the process described by equation (2.23), two singlet excitons are lost while in the process described by equation (2.24), only one exciton is lost per interaction. Hence, the effective rate coefficient for SSA is $2\cdot(1-\xi)\kappa_{SS}+\xi\kappa_{SS}=(2-\xi)\kappa_{SS}$. The second term is the contribution of STA. The rate term is proportional to the singlet exciton density as well as the density of the triplet excitons. As shown in equation (2.32), each singlet-triplet interaction annihilates one singlet exciton. The third term is the number of singlet excitons created by bimolecular triplet-triplet annihilation, as described in equation (2.36). The contributions of SPA and ISC processes are given by the last two terms in equation (3.23). Thus, most annihilation processes reduce the singlet exciton density, which means that they reduce the population of the upper laser level.

3.4.1.3 Triplet exciton rate equation for BA

The triplet exciton rate term for annihilation processes is given by:

$$
\left.\frac{\partial n_{T_1}(x,t)}{\partial t}\right|_{BA} = +(1-\xi)\cdot\kappa_{SSA}n_{S_1}^2
$$
$$
-(1+\xi)\cdot\kappa_{TTA}n_{T_1}^2
$$
$$
-\kappa_{TPA}(n_e + n_h)n_{T_1} + \kappa_{ISC}n_{S_1}. \tag{3.24}
$$

The first term on the right hand side describes the number of triplet excitons produced by SSA. As shown in equation (2.23), one triplet exciton is created per interaction. The triplet exciton loss from TTA is expressed by the second term. The rate term is proportional to the square of the triplet exciton density, because TTA is a bimolecular process. In equation (2.35), one triplet exciton is lost per reaction while in equation (2.36), two triplets are annihilated. Hence, the effective rate coefficient for TTA is $(1-\xi)\kappa_{TT}+2\cdot\xi\kappa_{TT}=(1+\xi)\kappa_{TT}$. The third term describes TPA. ISC process is taken into account by the last term.

3.4.1.4 Summary

In our model, bimolecular annihilation processes are considered, which greatly reduce the quantum efficiency in organic semiconductor devices under high excitation. The impact of each annihilation process is determined by a rate coefficient. Unfortunately, none of the potential materials for the emission layer have been completely characterized in terms of annihilation processes. Therefore, annihilation parameters, which are in the range of typical small molecule OLED materials, have been employed as a starting point for the simulations. Implications for the entire range of material parameters can then be achieved by a parameter variation study.

3.4.2 Field-induced exciton dissociation

For organic semiconductors, electric fields of the order of MV/cm are necessary in order to drive current densities above 100 A/cm^2. Therefore, the additional process *field-induced exciton dissociation*, which is often referred to as *field quenching* (FQ) greatly influences the device performance under high excitation.

In organic semiconductors, the singlet exciton binding energy is in the range of 0.2 eV to about 1.0 eV whereas triplet exciton binding energies are reported in the range from 0.4 eV to about 1.6 eV [108–119]. Hence, excitons are stable at room temperature. In the absence of an electric field, the dominating deactivation channels are radiative and nonradiative decay, while the exciton dissociation rate to free charge carriers is rather small. When a strong electric field is applied, the exciton dissociation rate is greatly increased. Depending on the exciton binding energies, field induced exciton dissociation becomes an important loss mechanism for electric fields above 10^6 V/cm [107]. Here, lower dissociation rates require higher exciton binding energies.

Excitons are dissociated to separated charge carriers. Hence, FQ of singlet and triplet excitons is described by the following reaction equations

$$ \mathrm{S_1} \xrightarrow{k_{\mathrm{FQ}}^{\mathrm{S_1}}} \mathrm{e+h} \; , \tag{3.25} $$

$$ \mathrm{T_1} \xrightarrow{k_{\mathrm{FQ}}^{\mathrm{T_1}}} \mathrm{e+h} \; . \tag{3.26} $$

Hence, the FQ induced change of the polaron densities n_{e} and n_{h} can be calculated from the change of the exciton densities:

$$ \left. \frac{\partial n_{\mathrm{e,h}}(x,t)}{\partial t} \right|_{\mathrm{FQ}} = - \left. \frac{\partial n_{\mathrm{S_1}}(x,t)}{\partial t} \right|_{\mathrm{FQ}} - \left. \frac{\partial n_{\mathrm{T_1}}(x,t)}{\partial t} \right|_{\mathrm{FQ}} . \tag{3.27} $$

The dissociation rate of excitons strongly depends on the electric fields. However, the total value of the electric field is only present in the direction of the electric field. For a three-dimensional situation, the directional dependence of the electric field has to be considered. For an arbitrary direction, only the electric field component for this direction influences the dissociation probability of the exciton. Therefore, the effective dissociation probability has to be treated with a three-dimensional approach.

For polymers, several field-induced measurements have shown, that the experimental data rather agrees with one-dimensional dissociation models than three-dimensional models. This behavior can be understood by considering the anisotropy of polymers. Since polymer chains are often oriented in parallel to each other, a one-dimensional behavior can be found.

In this section, three models are presented, which have been employed successfully for describing experimental data of field-induced exciton quenching: the Poole-Frenkel [213], the Onsager theory [120] and the hopping separation (HS) model [130]. For all theories, satisfying agreement with measurements can be found [123, 124]. In addition, the need for considering the dynamics of exciton relaxation has been demonstrated [117, 128, 129], suggesting that exciton dissociation might not be described solely by a rate equation.

3.4.2.1 1D Poole-Frenkel theory

Here, the Poole-Frenkel theory [213] is described, which takes into account the quantum mechanical nature of organic semiconductors. The Poole-Frenkel theory describes the dissociation probability of a bound quantum state [214]. The rate equation for exciton dissociation to free charge carriers as a function of the electric field $F(x, t)$ is

$$\left. \frac{\partial n_{S_1,T_1}(x,t)}{\partial t} \right|_{\mathrm{FQ,PF1D}} = -\frac{\nu}{\exp\left(\frac{E_{0,S_1,T_1} - \beta F(x,t)^{0.5}}{k_B T}\right)} n_{S_1,T_1}(x,t) \tag{3.28}$$

$$\left. \frac{\partial n_{e,h}(x,t)}{\partial t} \right|_{\mathrm{FQ,PF1D}} = -\left. \frac{\partial n_{S_1}(x,t)}{\partial t} \right|_{\mathrm{FQ,PF1D}} - \left. \frac{\partial n_{T_1}(x,t)}{\partial t} \right|_{\mathrm{FQ,PF1D}} \tag{3.29}$$

where β is defined by

$$\beta = \left(\frac{e^3}{\epsilon_0 \epsilon_r}\right)^{1/2} \tag{3.30}$$

and ν is

$$\nu = \frac{D_{S_1,T_1} \epsilon_0 \epsilon_r F(x,t)}{e}. \tag{3.31}$$

Here, D_{S_1} and D_{T_1} are the diffusion constants of the singlet and triplet excitons,

ϵ_0 denotes the dielectric constant of vacuum, ϵ_r is the relative dielectric constant and e stands for the elementary charge. The material parameters exhibiting the most striking impact on the dissociation rates are the binding energies of singlet and triplet excitons, denoted E_{0,S_1} and E_{0,T_1}, respectively. The Boltzmann constant is denoted k_B and T is the temperature.

3.4.2.2 3D Poole-Frenkel theory

The three-dimensional Poole-Frenkel effect has been treated by Hartke [215], who considered the escape rate of an electron, which is bound to a positive point charge. The field-dependent dissociation rate is

$$\left. \frac{\partial n_{S_1,T_1}(x,t)}{\partial t} \right|_{FQ,PF3D} = -\, k_{FQ,PF3D,F=0} \times \left[\left(\frac{k_B T}{\beta \sqrt{F(x,t)}} \right)^2 \right.$$
$$\left. \times \left\{ 1 + \left[\left(\frac{\beta \sqrt{F(x,t)}}{k_B T} \right) - 1 \right] \exp \left(\frac{\beta \sqrt{F(x,t)}}{k_B T} \right) \right\} + \frac{1}{2} \right]$$

where the dissociation rate at zero field $k_{FQ,PF3D,F=0}$ is

$$k_{FQ,PF3D,F=0} = \nu \exp \left(-\frac{E_{0,S_1,T_1}}{k_B T} \right), \tag{3.32}$$

and

$$\beta = \left(\frac{e^3}{\pi \epsilon_0 \epsilon_r} \right)^{1/2}. \tag{3.33}$$

ν has been defined in equation 3.31.

3.4.2.3 Onsager theory

Exciton dissociation in amorphous materials is often described by the Onsager theory [120], which has been extended by Braun [121] and Goliber [122]. For this theory, satisfying agreement with measurements has been achieved [123–127]. However, the need for considering the dynamics of exciton relaxation has been demonstrated [117, 128, 129], suggesting that exciton dissociation might not be described solely by a rate equation. Besides the Onsager theory, further models describing the impact of FQ have been discussed in literature [130, 131]. However, the Onsager theory seems to describe the effect of FQ correctly for most materials [125, 130, 131].

Experimental values for field-induced photoluminescence quenching are only available up to electric fields of up to 4 MV/cm [107, 133]. However, for the considered device and material properties, it turns out in our work that local fields up to 20 MV/cm would be necessary in order to achieve laser threshold. We note that this is well above the electrical breakdown field of typical thin film organic semiconductor devices.

Here, the theory presented in [121] and [122] is employed in order to describe the field dependent dissociation rate of excitons. The rate equation for exciton dissociation into a pair of free charge carriers (field quenching) as a function of the electric field $F(x,t)$ is

$$\left.\frac{\partial n_{S_1,T_1}(x,t)}{\partial t}\right|_{\text{FQ,ONS}} = -\frac{e(\mu_e + \mu_h)}{2\epsilon_0\epsilon_r}\frac{3}{4\pi a^3}\exp\left(-\frac{E_{0,S_1,T_1}}{k_B T}\right)$$
$$\times \frac{J_1\left[2\sqrt{2}\,(-b)^{0.5}\right]}{\sqrt{2}\,(-b)\,0.5} n_{S_1,T_1}(x,t). \quad (3.34)$$

$$\left.\frac{\partial n_{e,h}(x,t)}{\partial t}\right|_{\text{FQ,ONS}} = -\left.\frac{\partial n_{S_1}(x,t)}{\partial t}\right|_{\text{FQ,ONS}} - \left.\frac{\partial n_{T_1}(x,t)}{\partial t}\right|_{\text{FQ,ONS}}. \quad (3.35)$$

Here, J_1 is the first order Bessel function, which is computed from the following power series (for b < 3) [120]

$$\frac{J_1\left[2\sqrt{2}\,(-b)^{0.5}\right]}{\sqrt{2}\,(-b)\,0.5} = 1 + b + \frac{b^2}{3} + \frac{b^3}{18} + \frac{b^4}{180} \quad (3.36)$$
$$+ \frac{b^5}{2700} + \frac{b^6}{56700} + \dots \quad (3.37)$$

For fields above 5×10^4 V/m, b becomes larger than 3. In this case, the following asymptotic expansion is employed [120]

$$\frac{J_1\left[2\sqrt{2}\,(-b)^{0.5}\right]}{\sqrt{2}\,(-b)\,0.5} = \left(\frac{2}{\pi}\right)^{0.5}(8b)^{-0.75}\exp\left((8b)^{0.5}\right)$$
$$\times \left\{1 - \frac{3}{8\,(8b)^{0.5}} - \frac{15}{128.8b} - \frac{105}{1024\left((8b)^{0.5}\right)^3} - \dots\right\}. \quad (3.38)$$

ϵ_0 denotes the dielectric constant of vacuum, ϵ_r is the relative dielectric constant and e stands for the elementary charge. μ_e and μ_h are the charge carrier mobilities of electrons and holes. The material parameters exhibiting the most striking impact on the dissociation rates are the binding energies of singlet and triplet excitons, denoted E_{0,S_1} and E_{0,T_1}, respectively. The Boltzmann constant is denoted k_B and T is the temperature. The parameter b is given by

$$b = \frac{e^3 F}{8\pi\epsilon_0\epsilon_{\mathrm{r}}k_{\mathrm{B}}^2 T^2}. \tag{3.39}$$

Assuming Coulomb binding, the initial separation distance a is

$$a = \frac{e^2}{4\pi\epsilon_0\epsilon_{\mathrm{r}} E_{0,\mathrm{S}_1,\mathrm{T}_1}}. \tag{3.40}$$

3.4.2.4 Hopping separation model

The hopping separation (HS) model has been proposed for describing the impact of field-induced reduction of the luminescence efficiency in amorphous thin films [130]. For the field-dependent hopping separation of charges, the following rate equation is derived

$$\left.\frac{\partial n_{\mathrm{S}_1,\mathrm{T}_1}(x,t)}{\partial t}\right|_{\mathrm{FQ,HS}} = k_{\mathrm{FQ,HS},F=0} \times \exp\left(\frac{e\delta F}{k_{\mathrm{B}}T}\right). \tag{3.41}$$

Here, δ is the carrier hopping distance of the charge separation process. The parameters $k_{\mathrm{FQ,HS},F=0}$ and δ are obtained as fit parameters from measurement. For Alq$_3$, best agreement between simulation and experiment has been achieved for a hopping distance of $\delta = 0.39$ nm and a zero-field charge separation quantum efficiency of $\eta_{\mathrm{CT}}(F=0) = 0.1$, which is

$$\eta_{\mathrm{CT}}(F=0) = \frac{k_{\mathrm{FQ,HS},F=0}}{k_{\mathrm{FQ,HS},F=0} + \tau^{-1}}. \tag{3.42}$$

For $\tau = 14$ ns [63], the zero-field rate coefficient is $k_{\mathrm{FQ,HS},F=0} = 8 \times 10^6$ /s.

3.4.2.5 Material parameters

In this section, an overview of literature values for singlet and triplet exciton binding energies in organic semiconductors is given. Table 3.4 summarizes some published binding energies for different polymer and small molecule materials, which have been acquired by measurement or by theoretical investigation. A good overview of singlet exciton binding energies can be found in [110].

Fundamental parameters of the Poole-Frenkel and the Onsager model are the binding energies of the singlet and triplet excitons, which determine the inset of field-induced exciton dissociation. Typically, the singlet exciton binding energy E_{S_1} is in the range of 0.2 eV to 1.4 eV [109–112], which corresponds to initial separation distances in the range of 0.3 nm to about 1.5 nm. The triplet exciton binding energy E_{T_1} is about 0.2 eV up to 1.2 eV higher than the singlet exciton binding energy [109, 111].

Table 3.4: Literature values for binding energies of singlet and triplet excitons for various organic semiconductor materials.

Material	$E_{\mathrm{S_1,0}}$ /eV	$E_{\mathrm{T_1,0}}$ /eV	Ref.
Alq$_3$	1.4		[112, 113]
α-NPD	1.0		[112]
Anthracene	1.0		[114]
CuPc	0.6		[112]
MEH-PPV	0.35		[108, 115]
MeLPPP	0.68	1.18	[109]
p-AOPV-co-PPV	0.36		[108]
PDA	0.4		[116]
PFO	0.30		[108]
PPV	0.0-0.6	1.15, 1.86	[109, 111, 117, 118]
PTCDA	0.8		[108, 119]
Standard material	0.4	0.6	

For the Onsager model, satisfying agreement with experimental data has been achieved. Thus, for the investigation of organic laser diodes, the Onsager model is employed [123–127]. In this work, binding energies of $E_{\mathrm{S_1}} = 0.4$ eV for singlet excitons and $E_{\mathrm{T_1}} = 0.6$ eV for triplet excitons are assumed. The impact of the singlet exciton binding energy $E_{\mathrm{S_1}}$ and the energy splitting $E_{\mathrm{T_1}} - E_{\mathrm{S_1}}$ on the laser threshold current density is analyzed in section 6.4.2.

3.4.2.6 Summary

In this section, three models for field-induced exciton dissociation have been discussed. In literature, good agreement with experimental data has been achieved for the Onsager model. Thus, the Onsager model is employed for modelling the impact of field-induced exciton dissociation in organic laser diodes.

3.4.3 Induced absorption processes

In order to determine the impact of polaron and exciton absorption on the threshold current density, a model has been employed describing the effect of various particle densities on the attenuation of the laser mode. The model takes into account the spatial distribution of the particle densities and the modal intensity profile in order to calculate a device and material dependent dimensionless absorption coefficient ζ, describing the actual impact of induced absorption processes on the laser threshold.

3.4.3.1 Net modal gain

In optically pumped organic laser devices, the modal gain g_{SE} is generated by singlet excitons via stimulated emission. In the case of pulsed optical pumping, the density of polarons and, for pulsed operation, the density of triplet excitons is negligible, so there is no charge carrier or exciton induced absorption, hence the net gain g_net is equal to the gain generated by stimulated emission, g_SE.

When an organic laser device is excited electrically, the densities of polarons and triplet excitons exceed the singlet exciton density by about two orders of magnitude for typical OLED materials [216]. Hence, the gain g_SE caused by stimulated emission is reduced by several absorption processes. The resulting net gain, g_net, is

$$g_\mathrm{net} = g_\mathrm{SE} - \alpha_\mathrm{e} - \alpha_\mathrm{h} - \alpha_{\mathrm{T_1 T_N}}. \tag{3.43}$$

Charge carrier absorption (also known as polaron absorption) describes the process of a polaron being excited to a higher lying state from where it quickly relaxes back to thermal equilibrium. Polaron absorption by negatively and positively charged polarons is referred to as α_e and α_h, respectively. In inorganic devices, this process is denoted as free carrier absorption and its effect on laser operation has been studied extensively (see e. q. [217]).

Singlet and triplet excitons can also quench laser light whereby the excitons are excited to a higher lying state. While the excitation of a singlet exciton into a higher lying state can be taken into consideration by using the effective cross section for stimulated emission [99], the case of triplet excitons has to be treated in a more elaborate manner.

If a photon is absorbed by an exciton in the first excited triplet state $\mathrm{T_1}$, the triplet exciton is excited to a higher lying triplet state, from where it quickly relaxes to lower lying states. The attenuation coefficient for this particular processes is $\alpha_{\mathrm{T_1 T_N}}$.

3.4.3.2 Gain and attenuation coefficients

The amplification of the laser mode due to stimulated emission can be expressed as a function of the intensity profile of the laser mode $I(x)$ combined with the density of singlet excitons $n_{\mathrm{S_1}}(x)$ in the active layer. We assume that the device is in steady state, hence the particle densities do not depend on time. Thus, the modal gain is

$$g_\mathrm{SE} = \frac{\int_0^d \sigma_\mathrm{SE} I(x)\, n_{\mathrm{S_1}}(x)\, dx}{\int_0^d I(x)\, dx}, \tag{3.44}$$

where d is the device thickness, $I(x)$ describes the intensity profile of the laser mode and $n_{\mathrm{S_1}}(x)$ represents the density of singlet excitons in the active layer. The cross section for stimulated emission is σ_SE.

In analogy, the attenuation coefficient for polaron absorption can be deduced. Polaron absorption is proportional to the weighted spatially dependent polaron density, hence the attenuation coefficients for polaron absorption are

$$\alpha_\mathrm{e,h} = \frac{\int_0^d \sigma_\mathrm{e,h} I(x)\, n_\mathrm{e,h}(x)\, dx}{\int_0^d I(x)\, dx}, \tag{3.45}$$

whereby the density of negatively and positively charged polarons are $n_e(x)$ and $n_h(x)$, respectively. The cross sections for polaron absorption are $\sigma_e(x)$ and $\sigma_h(x)$, respectively.

For triplet-triplet absorption the attenuation coefficient is

$$\alpha_{T_1 T_N} = \frac{\int_0^d \sigma_{T_1 T_N} I(x)\, n_{T_1}(x)\, dx}{\int_0^d I(x)\, dx},$$
(3.46)

where the density of triplet excitons is $n_{T_1}(x)$. Since the transition $T_1 \rightarrow T_N$ is allowed, the lifetimes of the higher lying triplet states are much smaller than the lifetime of the T_1 state (which is typically of the order of 10^{-6} s to 10^{-3} s [218,219]). Thus, we assume that the T_N states are virtually unpopulated.

3.4.3.3 Definition of absorption parameters

In the previous section, we have introduced the modal gain g_{SE} due to stimulated emission and the attenuation coefficients α_e, α_h and $\alpha_{T_1 T_N}$ describing polaron and triplet-triplet absorption. These are determined by the cross sections σ for the individual processes, the intensity profile $I(x)$ of the laser mode and the spatial distribution of the densities $n(x)$ of the involved particles. While the cross sections σ are inherent material properties, $I(x)$ as well as $n(x)$ also depend on the device design and the mode of operation. In order to quantify the role of $I(x)$ and $n(x)$ for polaron and triplet-triplet absorption we introduce a dimensionless absorption parameter ζ describing the relative amount of polarons and triplet excitons (weighted with the intensity profile) compared to singlet excitons.

For polaron absorption the absorption parameters are

$$\zeta_{e,h} = \frac{\int_0^d I(x)\, n_{e,h}(x)\, dx}{\int_0^d I(x)\, n_{S_1}(x)\, dx}$$
(3.47)

Here ζ_e and ζ_h describe negatively and positively charged polarons. The corresponding absorption parameter for triplet-triplet absorption is

$$\zeta_{T_1} = \frac{\int_0^d I(x)\, n_{T_1}(x)\, dx}{\int_0^d I(x)\, n_{S_1}(x)\, dx}.$$
(3.48)

Using equation (3.47) and equation (3.48), equation (3.43) can be rewritten as

$$g_{net} = g_{SE} \left(1 - \zeta_e \frac{\sigma_e}{\sigma_{SE}} - \zeta_h \frac{\sigma_h}{\sigma_{SE}} - \zeta_{T_1} \frac{\sigma_{T_1 T_N}}{\sigma_{SE}} \right).$$
(3.49)

For simplicity, we assume that the absorption cross sections σ_e and σ_h for electrons and holes are equal. Defining $\sigma_{carrier} = \sigma_e = \sigma_h$ and $\zeta_{carrier} = \zeta_e + \zeta_h$, we obtain

$$g_{net} = g_{SE} \left(1 - \zeta_{carrier} \Sigma_{carrier} - \zeta_{T_1} \Sigma_{T_1 T_N} \right),$$
(3.50)

where the dimensionless relative absorption cross sections $\Sigma_{\text{carrier}} = \sigma_{\text{carrier}}/\sigma_{\text{SE}}$ and $\Sigma_{\text{T}_1\text{T}_\text{N}} = \sigma_{\text{T}_1\text{T}_\text{N}}/\sigma_{\text{SE}}$ are defined. For laser operation, the net gain g_{net} has to be positive. This implies that the following condition has to be fulfilled

$$\zeta_{\text{carrier}}\Sigma_{\text{carrier}} + \zeta_{\text{T}_1}\Sigma_{\text{T}_1\text{T}_\text{N}} < 1. \tag{3.51}$$

From equation (3.51) necessary but not sufficient conditions can be deduced for the different species

$$\zeta_{\text{carrier}} < \Sigma_{\text{carrier}}^{-1} \tag{3.52}$$

$$\zeta_{\text{T}_1} < \Sigma_{\text{T}_1\text{T}_\text{N}}^{-1}. \tag{3.53}$$

Since the relative absorption cross sections Σ_{carrier} and $\Sigma_{\text{T}_1\text{T}_\text{N}}$ are material constants, ζ_{carrier} and ζ_{T_1} can be used to compare different laser devices concerning their polaron and triplet-triplet absorption characteristics.

3.4.4 Summary

In this section, the modelling of loss processes in organic laser diodes has been discussed. Due to the low charge carrier mobilities of organic semiconducting materials, both charge carrier densities of about 10^{19} /cm³ and electric fields of the order of MV/cm are required to achieve current densities of the order of 1 kA/cm². Thus, loss processes, which are caused by high electric fields and high particle densities have to be considered. In our model, the impact of bimolecular annihilations is included by reaction rates. As a starting point for our calculations, rate coefficients of typical small molecule materials are employed. The impact of field quenching is described by an Onsager approximation. Regarding induced absorption losses, only very few absorption cross sections have been published, yet. Therefore, necessary conditions for the cross sections for polaron and triplet absorption have been derived, which have to be fulfilled in order to achieve laser operation.

3.5 Chapter summary

In this work, an integrated model of the electrical and optical properties of an organic laser structure is employed in order to study the impact of various loss processes on the laser threshold current density. The dynamics of electrons, holes and singlet excitons are treated in the framework of a self consistent numerical drift-diffusion model. This model of the electrical properties provides the particle distributions to the optical model. The actual waveguide structure is considered and the resulting laser rate equations are solved.

4 Impact of bimolecular annihilation processes

In this chapter, the electrical and optical features of TE_0-mode organic double-heterostructure laser diodes are analyzed by numerical simulation. A potential organic laser diode structure has to combine low waveguide losses, good charge carrier injection and a high quantum efficiency. For low waveguide losses, the self-absorption of the organic layers has to be low. Since the energy scheme of organic semiconducting materials corresponds to a four-level situation, this requirement is fulfilled and optically excited lasing can be achieved.

For electrically pumped lasing, highly conductive electrodes are required, which support current densities of the order of 1 kA/cm^2. However, highly conductive electrodes are in general strongly absorbing, thus the attenuation of the laser waveguide is increased. In order to reduce the waveguide attenuation, the laser mode has to be separated from the emission layer of the device and low-absorption electrodes have to be employed. For excellent electrical properties, the injection barriers for the charge carriers have to be small. In addition, the transport layers have to block oppositely charged carriers in order to achieve high quantum efficiencies.

In this chapter, a device structure is discussed, which is based on the double-heterostructure OLED design for propagation of a TE_0 laser mode and which fulfills many of the abovementioned requirements. The impact of bimolecular annihilation processes in organic laser diodes is analyzed. First, the device geometry of the considered device is discussed in section 4.1. Then, the impact of bimolecular annihilation processes on the threshold current density is studied in section 4.2. Section 4.3 discusses the impact of each annihilation process on the threshold current density. The results of this chapter are summarized in section 4.4.

4.1 Device geometry of organic TE_0-mode laser diode

The investigated three-layer laser diode structure is shown in figure 4.1. The considered device consists of an organic layer system with two transport layers and one emission layer, electrodes and a silicon substrate. A waveguide for the laser mode is formed by the organic layers. It has been shown by simulation that the quality of this waveguide is of crucial importance for laser operation [164].

The employed mobilities of electrons and holes for the respective organic layers are

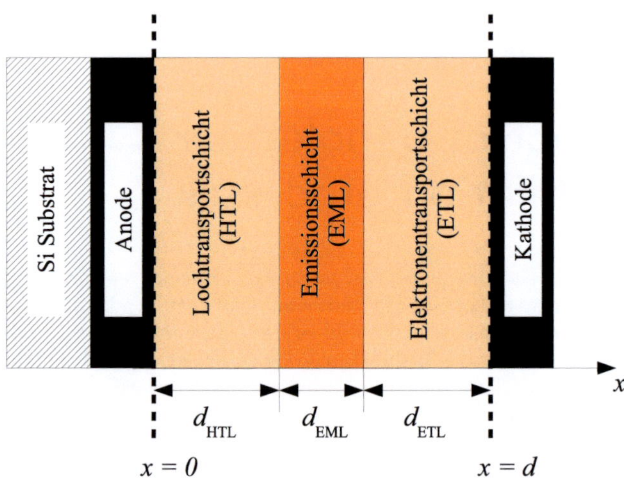

Figure 4.1: Structure of the simulated three layer laser diode with hole transport (HTL), emission (EML) and electron transport layer (ETL).

in the range of typical materials used for OLED fabrication. For the transport layers, a mobility of 5×10^{-4} cm^2/Vs has been assumed for the respective charge carrier. In the emission layer, the mobility of both charge carriers is assumed to be two orders of magnitude lower than the mobility in the transport layers. Electrons are blocked by the HTL while holes are blocked by the ETL. Due to this, the emission layer thickness is an important parameter for annihilation processes. The refractive index of both transport layers is assumed to be equal.

A low attenuation coefficient of the waveguide has been found to be inevitable for laser operation, which has also been demonstrated experimentally for optically pumped devices with highly conductive electrodes [165]. In our model, the organic layers are assumed to be transparent while the electrodes exhibit a high absorption coefficient of 10^6 cm^{-1}, which is a typical value for metals. Hence, the attenuation coefficient of the waveguide at the laser wavelength is determined by the confinement factor of the laser mode in the electrode layers. By increasing the thickness of the transport layers from 100 nm to 600 nm, the confinement factor of the laser mode in the electrode layers is decreased, which means that the attenuation coefficient can be reduced by two orders of magnitude. By further increasing the transport layer thickness from 600 nm to 1000 nm, the attenuation coefficient can be reduced by another factor of three, but a higher voltage is required for a specific current density. Hence, a total device thickness of 1300 nm is used as a trade-off between the optical and electrical performance of the device.

4.2 Impact of annihilation processes on the laser threshold current density

In figure 4.2, the calculated threshold current density is shown for different emission layer thicknesses while the total device thickness is kept constant. The dashed curve displays the threshold current density when annihilation processes are neglected. In this case, the threshold current density is $j_{\mathrm{thr}} = 170$ A/cm^2 (voltage at laser threshold is $U_{\mathrm{thr}} = 1.95$ kV) for the optimum emission layer thickness of 400 nm. If the emission layer thickness is reduced to 200 nm the threshold current density increases by a factor of 1.3 to $j_{\mathrm{thr}} = 214$ A/cm^2 ($U_{\mathrm{thr}} = 1.42$ kV).

When annihilation processes are included, the impact of the emission layer thickness is much more pronounced. The solid curve in figure 4.2 shows the threshold current density when annihilation processes are present. At $d_{\mathrm{EML}} = 400$ nm, the threshold current density is $j_{\mathrm{thr}} = 564$ A/cm^2 ($U_{\mathrm{thr}} = 3.53$ kV). If the emission layer thickness is reduced to $d_{\mathrm{EML}} = 200$ nm, the threshold current density increases strongly to $j_{\mathrm{thr}} = 1334$ A/cm^2 ($U_{\mathrm{thr}} = 3.51$ kV). The voltage at the laser threshold also increases considerably, if annihilation processes are included. For the optimum emission layer thickness, the required voltage increases from 1.42 kV to 3.51 kV. Hence, high voltage sources are required for the operation of the considered device.

The distinctive influence of the emission layer thickness on the threshold current density arises from the nature of bimolecular annihilation processes. If the emission layer thickness is reduced, the attenuation coefficient of the laser mode slightly increases. Hence, a higher singlet exciton density is required to reach laser threshold. This explains the slight increase of threshold current density for the case without annihilation processes. If annihilation processes are included, an increase of the exciton density leads to a strong increase of the annihilation rate. Hence, the influence of the emission layer thickness is much more pronounced if annihilation processes are included. This qualitative explanation can be quantified: In figure 4.3, the total exciton annihilation rate density is plotted against the position in the laser device at laser threshold. The total loss rate includes the exciton loss rates due to SSA, STA and SPA. By increasing the emission layer thickness from 100 nm to 500 nm the annihilation rate density can be reduced by a factor of 100. At the edge of the emission layer the annihilation rate density is much higher than in the middle. This effect results from the assumption that polarons are blocked by the opposite transport layer causing an accumulation of polarons at the interface of the emission layer and the opposite transport layer. The pronounced dependence of the annihilation density on the emission layer thickness shows the fundamental importance of a careful optimization of the device dimensions and demonstrates that thin emission layers are not preferable in the presence of annihilation processes.

In figure 4.4, the contributions of the different quenching processes are shown as a function of the emission layer width at the laser threshold current density. The ratios of the annihilation processes only slightly depend on the emission layer width between $d_{\mathrm{EML}} = 200$ nm and $d_{\mathrm{EML}} = 700$ nm. At an emission layer width of 400 nm,

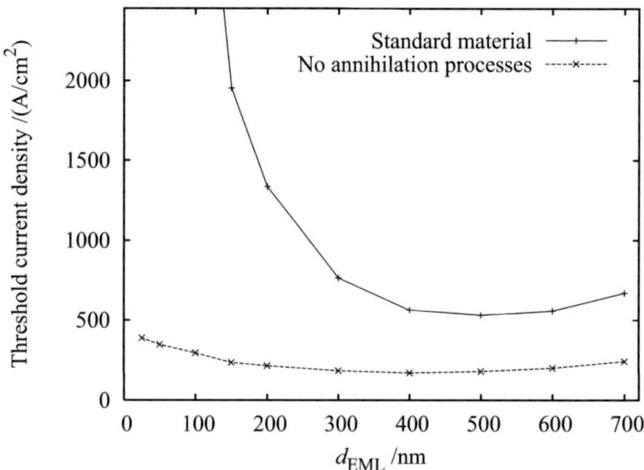

Figure 4.2: Threshold current density as a function of the emission layer thickness with annihilation processes and in their absence. The total device thickness is 1300 nm.

74% of the lost singlet excitons are quenched by SPA and about 26% are lost by STA. If the emission layer thickness is less than 200 nm, the importance of STA increases. At an emission layer thickness of 50 nm, 68% of the lost singlet excitons are quenched by SPA and about 32% are lost by STA. The singlet exciton density is not reduced significantly by SSA.

In the following subsections, the influence of a parameter variation of the annihilation rate coefficients on the threshold current density is examined. Only one parameter is varied at a time. The standard parameter set (see table 3.3) is used as a starting point for our parameter variation study. First, the annihilation rate coefficient under investigation is set to zero in order to simulate the device operation without the considered process. Then, the annihilation rate coefficient is increased to calculate the maximum value for the rate coefficient to allow a threshold current density below 1000 A/cm^2 for the optimum emission layer thickness. This study can be employed as a guideline for device design and to evaluate the potential of different organic semiconducting materials for use in electrically pumped organic lasers.

4.3 Bimolecular annihilations

The annihilation rate coefficients can differ by up to three orders of magnitude for different organic semiconducting materials. Most of them are still unknown and no material has been completely characterized yet. In this study, a set of annihilation

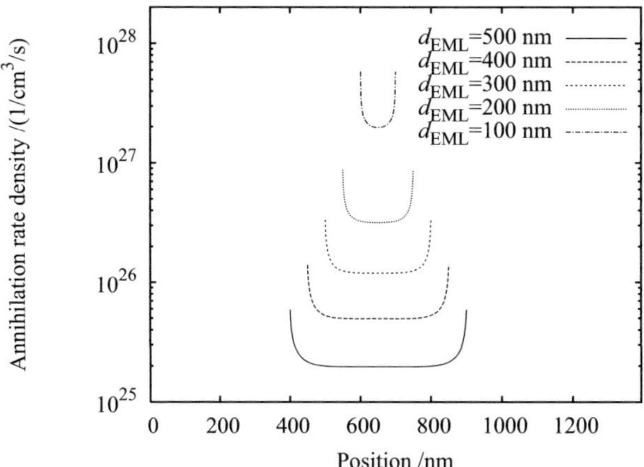

Figure 4.3: Total S_1 annihilation rate density as a function of position in the device for different emission layer thicknesses.

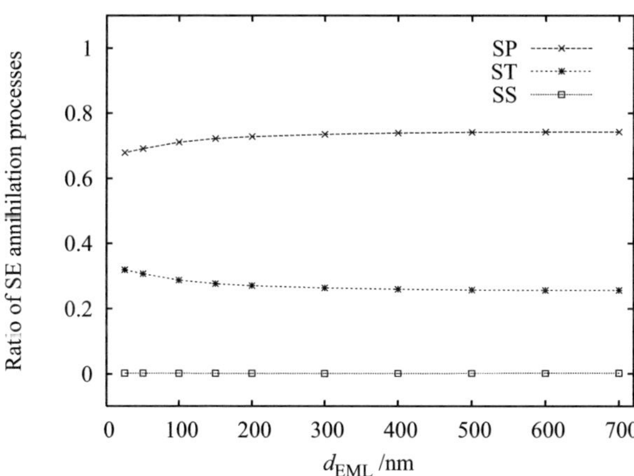

Figure 4.4: Contribution of the various S_1 annihilation processes to the total S_1 annihilation rate.

constants is employed which are in the range of rate coefficients for typical organic materials (see table 3.3). They are not meant to describe a specific material, but are rather used as a starting point for our numerical study. By systematically varying the rate coefficients, the role of each annihilation process is examined. Any dependence of the annihilation parameters on other parameters, such as the mobility of electrons and holes or the diffusion constant of excitons, is ignored. The results of this parametric study can be used as guidelines for material selection and device development.

4.3.1 Singlet-singlet annihilation (SSA)

In this subsection, the influence of SSA on the threshold current density is examined. As discussed in the preceding section, SSA does not significantly quench the singlet excitons when standard parameters are used. In figure 4.5, the threshold current density is plotted as a function of the emission layer thickness for different SSA rate coefficients. The solid line shows the threshold current density for the standard parameters. The corresponding rate coefficient is indicated in the key of the figure with an asterisk. The threshold current density is omitted for emission layer thicknesses below $d_{\mathrm{EML}} = 200$ nm, because the threshold current density exceeds 4000 A/cm^2 in this case.

If κ_{SS} is changed from its standard value of $\kappa_{\mathrm{SS}} = 10^{-10}$ cm^3/s to zero, the threshold current density does not change noticeably. In figure 4.5, the curves for the two cases are virtually identical. Hence the threshold current density is not increased significantly by SSA with standard parameters. If the rate coefficient is increased by a factor of 100 to $\kappa_{\mathrm{SS}} = 10^{-8}$ cm^3/s, the minimum threshold current density increases from $j_{\mathrm{thr}} = 564$ A/cm^2 to $j_{\mathrm{thr}} = 662$ A/cm^2. A further increase of the rate coefficient increases the threshold current density abruptly. At an SSA rate coefficient of $\kappa_{\mathrm{SS}} = 5 \times 10^{-8}$ cm^3/s, we find a threshold current density of $j_{\mathrm{thr}} = 1031$ A/cm^2. Thus, SSA rate coefficients up to $\kappa_{\mathrm{SS}} \approx 5 \times 10^{-8}$ cm^3/s might allow laser operation with a threshold current density of the order of 1000 A/cm^2. The influence of the rate coefficients on the laser threshold is displayed in figure 4.6, where the threshold current density is shown as a function of different rate coefficients for a fixed emission layer thickness of $d_{\mathrm{EML}} = 400$ nm.

4.3.2 Singlet-triplet annihilation (STA)

In figure 4.7, the threshold current density is shown as a function of the emission layer thickness for different STA rate coefficients. If the STA rate coefficient is changed from its standard value of $\kappa_{\mathrm{ST}} = 10^{-10}$ cm^3/s to zero, the minimum threshold current density decreases from $j_{\mathrm{thr}} = 564$ A/cm^2 to $j_{\mathrm{thr}} = 401$ A/cm^2. This illustrates the great influence of STA on the threshold current density. If the rate coefficient is increased by a factor of 3 to $\kappa_{\mathrm{ST}} = 3 \times 10^{-10}$ cm^3/s, the threshold current density increases considerably to $j_{\mathrm{thr}} = 1015$ A/cm^2 (see figure 4.6). This result demonstrates that STA is an important loss channel for singlet excitons and one of the major problems for electrically pumped organic lasers.

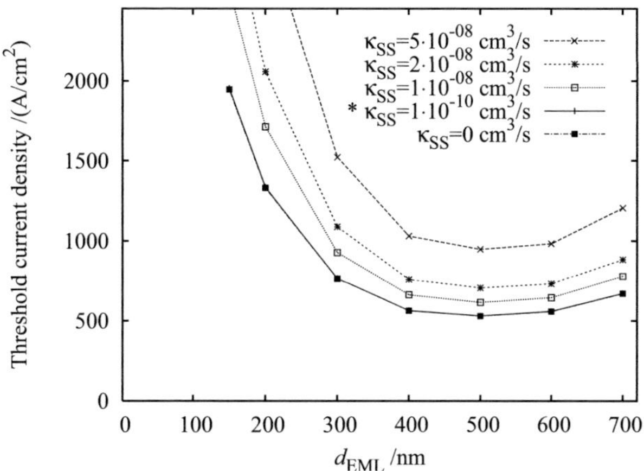

Figure 4.5: Threshold current density as a function of emission layer thickness for different singlet-singlet annihilation rate coefficients. The standard rate coefficient for SSA is indicated in the key of the figure with an asterisk. The curves for $\kappa_{SS} = 10^{-10}$ cm^3/s and for $\kappa_{SS} = 0$ are on top of each other.

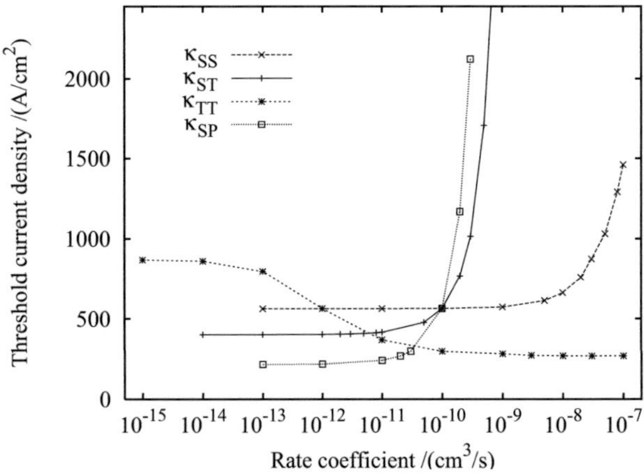

Figure 4.6: Threshold current density as a function of the rate coefficient for different annihilation processes at a constant emission layer thickness of $d_{\text{EML}} = 400$ nm.

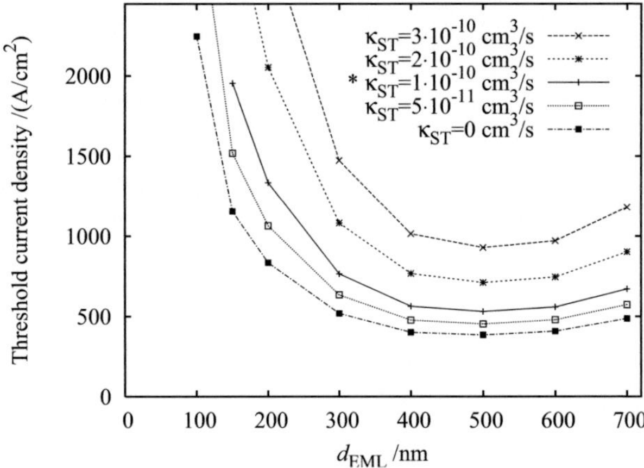

Figure 4.7: Threshold current density as a function of emission layer thickness for different singlet-triplet annihilation rate coefficients.

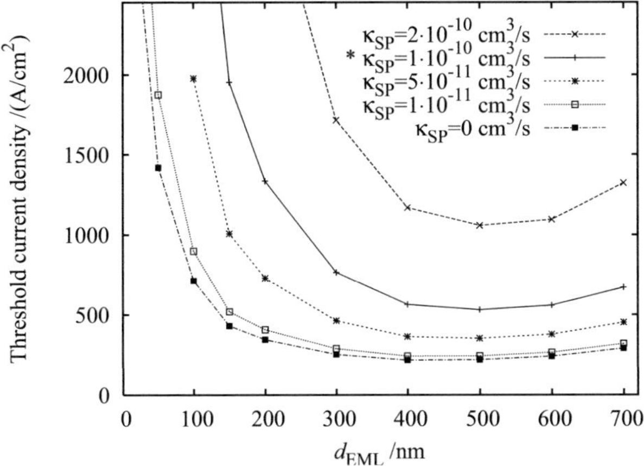

Figure 4.8: Threshold current density as a function of emission layer thickness for different singlet-polaron annihilation rate coefficients.

4.3.3 Singlet-polaron annihilation (SPA)

As discussed in section 4.2, the dominating loss channel for singlet excitons is SPA if the standard parameter set is used. In figure 4.8, the threshold current density is presented for several SPA rate coefficients as a function of the emission layer thickness. If the rate coefficient is changed from its standard value of $\kappa_{\mathrm{SP}} = 10^{-10}$ cm^3/s to zero, the laser threshold drops from $j_{\mathrm{thr}} = 564$ A/cm^2 to $j_{\mathrm{thr}} = 216$ A/cm^2, which demonstrates that SPA is the dominating annihilation process for the standard parameter set. If the SPA rate coefficient is increased to $\kappa_{\mathrm{SP}} = 2 \times 10^{-10}$ cm^3/s, the threshold current density increases very strongly to $j_{\mathrm{thr}} = 1169$ A/cm^2. This behavior is clarified in figure 4.6. These results demonstrate that the SPA rate coefficient should not be larger than $\kappa_{\mathrm{SP}} \approx 2 \times 10^{-10}$ cm^3/s in order to obtain a threshold current density of less than 1000 A/cm^2.

4.3.4 Triplet-triplet annihilation (TTA)

The threshold current density is also influenced by triplet exciton annihilation processes. On the one hand, TTA generates singlet excitons, on the other hand singlet excitons are quenched by triplets through STA. Both are bimolecular processes. The TTA rate is proportional to $n_{\mathrm{T}_1}^2$ and the STA rate is proportional to $n_{\mathrm{T}_1} n_{\mathrm{S}_1}$. Hence these processes become more important at high current densities.

In figure 4.9, the threshold current density is plotted as a function of the emission layer thickness for several TTA rate coefficients. If the TTA rate coefficient is changed

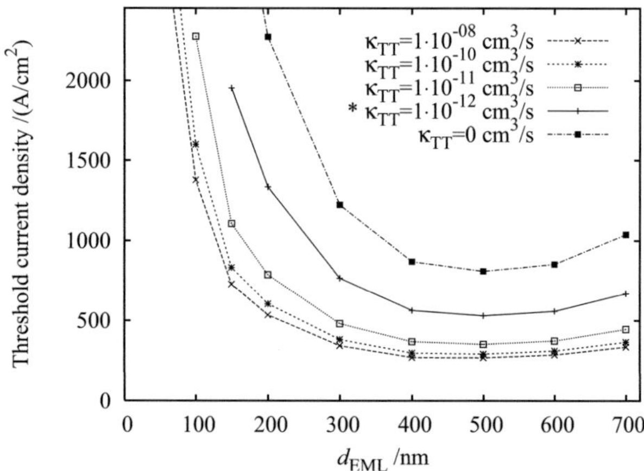

Figure 4.9: Threshold current density as a function of emission layer thickness for different triplet-triplet annihilation rate coefficients.

from its standard value of $\kappa_{TT} = 10^{-12}$ cm^3/s to zero, the threshold current density is increased from $j_{\text{thr}} = 564$ A/cm^2 to $j_{\text{thr}} = 870$ A/cm^2. If, on the other hand, the TTA rate coefficient is increased by a factor of 10 to $\kappa_{TT} = 10^{-11}$ cm^3/s the threshold current density decreases significantly from $j_{\text{thr}} = 564$ A/cm^2 to $j_{\text{thr}} = 369$ A/cm^2.

The former observation can be explained by two effects. First, a decrease of the TTA rate coefficient causes a decrease of the number of singlet excitons created by this reaction. Second, the triplet exciton density increases and more singlet excitons are annihilated by STA. An increased TTA rate reduces the triplet exciton density, which means that less singlet excitons are quenched by triplets. Additionally, there are more singlet excitons produced by TTA. If the TTA rate coefficient is increased further, the threshold current density saturates at $j_{\text{thr}} = 268$ A/cm^2 (see figure 4.6). In this case, the triplet density is virtually zero.

4.3.5 Triplet-polaron annihilation (TPA)

Figure 4.10 shows a plot of the threshold current density against the emission layer thickness for various triplet-polaron annihilation rate coefficients. If the TPA rate coefficient is changed from its standard value of $\kappa_{TP} = 10^{-12}$ cm^3/s to zero, the threshold current density decreases from $j_{\text{thr}} = 564$ A/cm^2 to $j_{\text{thr}} = 449$ A/cm^2. If the rate coefficient is increased by a factor of 10^4 to $\kappa_{TP} = 10^{-8}$ cm^3/s, the threshold current density decreases only slightly to $j_{\text{thr}} = 492$ A/cm^2.

If the TPA rate coefficient is varied over many orders of magnitude, the triplet exciton density is also altered by many orders of magnitude, but the threshold cur-

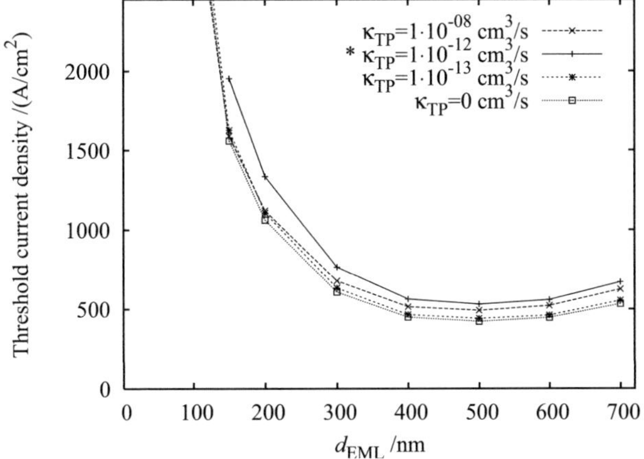

Figure 4.10: Threshold current density as a function of emission layer thickness for different triplet-polaron annihilation rate coefficients.

rent density only changes slightly. The triplet exciton density is reduced by TPA while the singlet exciton density is only affected indirectly. This behavior can be explained by considering the effects of singlet-triplet and triplet-triplet annihilation. TTA produces singlet excitons while STA eliminates singlet excitons. The difference between these two reaction rates represents the effective singlet exciton generation rate caused by triplets. In the next section, the influence of triplet excitons on the threshold current density is discussed in detail.

4.3.6 The role of triplet excitons

The singlet exciton density is decreased by triplet excitons via STA while TTA produces singlet excitons. In order to analyze the role of triplet excitons for laser operation, a *total effective rate* is examined as introduced in equation (4.1). It is defined as the total number of singlets excitons created and quenched by triplet excitons in steady state operation.

$$r_{\mathrm{total,eff}} = \int_0^d \left(\xi\kappa_{\mathrm{TT}}n_{\mathrm{T_1}}^2\left(x\right) - \kappa_{\mathrm{ST}}n_{\mathrm{S_1}}\left(x\right)n_{\mathrm{T_1}}\left(x\right) \right) dx \qquad (4.1)$$

In figure 4.11, the effective rate is plotted for different singlet-triplet annihilation constants as a function of current density. Employing standard parameters (solid curve), the effective rate is negative at low current densities, which means that slightly more singlet excitons are quenched by STA than created by TTA. At current densities

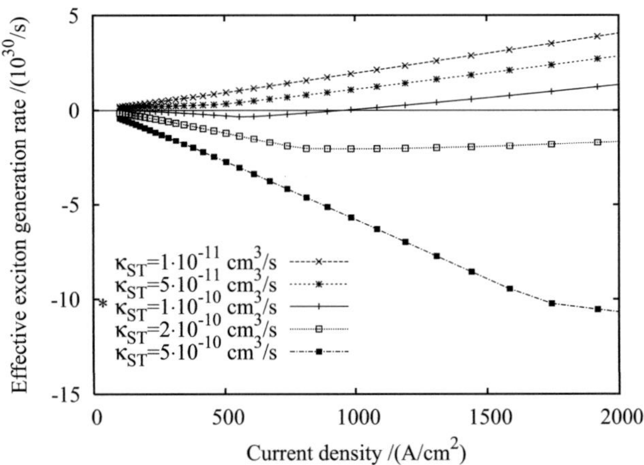

Figure 4.11: *Total effective rate* (see equation (4.1)) describing the number of singlet excitons created (or quenched) by triplet excitons in the entire device, shown as a function of current density for different STA rate coefficients.

of more than 1000 A/cm^2, the effective rate is positive, which means that more singlet excitons are created by triplet-triplet annihilation than quenched by singlet-triplet annihilation. If the singlet-triplet annihilation rate coefficient is increased, the singlet-triplet annihilation reaction dominates and the effective rate is negative for all current densities, hence the laser threshold current density is increased.

The absolute rate of STA is proportional to $n_{S_1} n_{T_1}$, whereas the absolute rate of TTA is proportional to $n_{T_1}^2$. For standard parameters, the triplet density is about 100 times higher than the singlet exciton density. Hence, the effective rate is nearly zero if $\kappa_{ST} = 100 \times \kappa_{TT}$. This effective rate depends on the particle density and on the rate coefficients of STA, TPA and TTA. The impact of the rate coefficients on the threshold current density is pronounced while the impact of the triplet exciton density on the threshold current density is rather small. The threshold current density increases strongly if $\kappa_{ST} \gg 100 \times \kappa_{TT}$.

4.4 Chapter summary

In this chapter, the influence of various annihilation processes on the threshold current density of a three layer organic laser structure has been investigated by numerical simulation for a TE$_0$-mode organic double-heterostructure device. Typical annihilation rates and material properties of organic semiconductors have been used as a starting point for the simulations. If annihilation processes are neglected, we obtain *a threshold current density of 170 A/cm² for an optimized design. Including annihilation processes increases the threshold current density to 564 A/cm² for the employed material parameters. The threshold current density is found to depend critically on the emission layer thickness. Singlet-polaron annihilation is identified as the dominating quenching process. Singlet-triplet annihilation has been recognized as another important annihilation process while the threshold current density is not increased significantly by singlet-singlet quenching and intersystem crossing.*

The rate coefficients of the standard material are applied as a starting point for a parameter variation study. Maximum rate coefficients are calculated which allow laser operation below 1 kA/cm², because such current densities are accessible with state-of-the-art technology and have been demonstrated in organic devices recently.

The singlet exciton density is influenced by triplet excitons via two annihilation processes: Singlet-triplet annihilation, which decreases the singlet exciton density and hence the available gain, and triplet-triplet annihilation, a process which creates singlet excitons. Thus, the net change in singlet excitons is influenced by the particle density and the rate coefficients. Our calculations show that the impact of the rate coefficients on the threshold current density is the dominating factor while the impact of the triplet exciton density is rather small.

5 Impact of polaron and triplet-triplet absorption

In organic semiconductors, charge carrier mobilities are low compared to their inorganic counterparts. Hence, high current densities come along with high particle densities. In the presence of bimolecular annihilations, the densities of polarons and triplet excitons exceed the singlet exciton density by about three orders of magnitude. Thus, induced absorptions by polarons and triplet excitons become important. In this chapter, the impact of polaron and triplet-triplet absorption is studied for the TE_0-mode device design, which has been discussed in section 4.1.

This chapter is structured as follows: In section 5.1, the polaron absorption parameter is calculated in the presence of bimolecular annihilations. In section 5.2, the dependencies of the polaron and the triplet absorption parameters on the current density are analyzed. In section 5.3, the laser threshold current density is calculated when polaron and triplet-triplet absorptions are included and the impact of the absorption parameters on the threshold current density is discussed. The impact of the device geometry and material parameters on the absorption parameters is studied in section 5.4 and in section 5.5. The transient characteristics of optical gain and the absorption parameters for pulsed operation is analyzed in section 5.6. The chapter is concluded with a brief summary in section 5.7.

5.1 Polaron absorption and bimolecular annihilations

In figure 5.1, the polaron absorption parameter ζ_{carrier} is plotted as a function of current density for the standard device. The solid line shows ζ_{carrier} for standard annihilation rate coefficients (see [216]) being used, whereas the dashed line shows the case, when all annihilation rate constants are set to zero.

Without annihilations, ζ_{carrier} follows the power law $j^{-0.5}$ for all shown current densities, which is a consequence of space charge limited current and Langevin recombination. Hence, for any set of cross sections for stimulated emission and polaron absorption, ζ_{carrier} can be reduced without limitations by increasing the pump power, hence laser operation can be achieved if the current density is high enough. The fact that ζ_{carrier} is decreasing for increasing current densities is caused by bimolecular Langevin type recombination of electrons and holes to excitons along with the monomolecular decay of the neutral excitons.

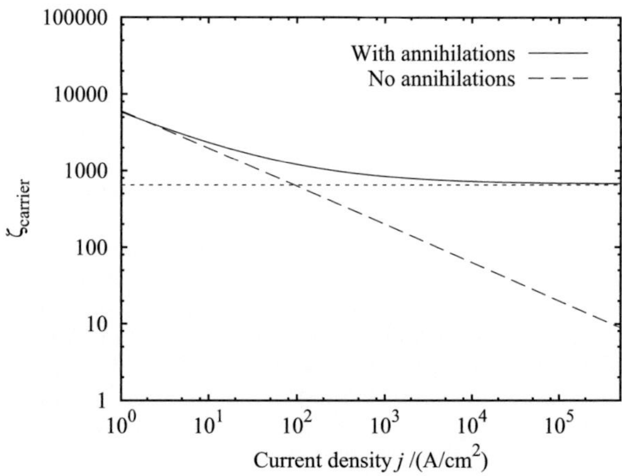

Figure 5.1: ζ_{carrier} as a function of current density with standard annihilations (solid line) and in their absence (dashed line). At current densities below $5\ \text{A/cm}^2$ the lines are virtually on top of each other. With annihilations, ζ_{carrier} is saturating at a value of 650 (dotted line) for current densities above $10^3\ \text{A/cm}^2$ while in the absence of annihilations, ζ_{carrier} follows the power $j^{-0.5}$ for all shown current densities.

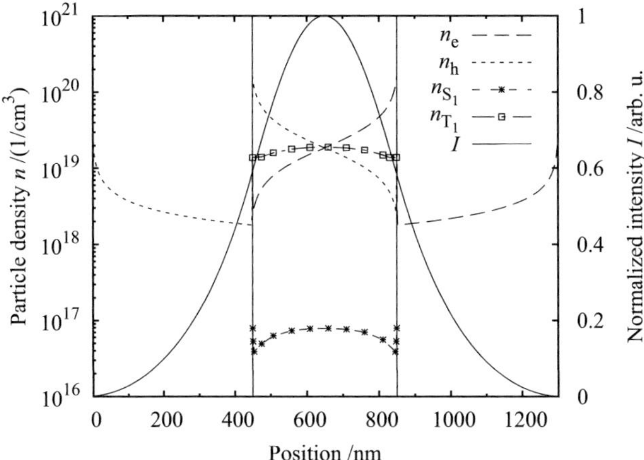

Figure 5.2: The spatial distribution of the polaron and exciton densities and the modal intensity profile as a function of the position in the device at a current density of 10^4 A/cm^2.

For annihilations being included, singlet excitons are also quenched by bimolecular annihilation processes like singlet-singlet and singlet-polaron annihilation. These annihilations become increasingly important at high current densities causing ζ_{carrier} to saturate at 650 for current densities above 10^3 A/cm^2.

5.2 Polaron and triplet absorption parameters

In this section we investigate devices with layer thicknesses of $d_{\text{HTL}} = d_{\text{ETL}} = 450$ nm, $d_{\text{EML}} = 400$ nm. Our calculations yield the spatially dependent particle densities as a function of the applied current density. The modal intensity profile I and the calculated distributions of carriers and excitons for a current density of 10^4 A/cm^2 are shown in figure 5.2. From this data the intensity profile weighted relative densities ζ_{carrier} and ζ_{T_1} are derived.

At the interface of hole transport layer and emission layer, the hole mobility μ_h drops by two orders of magnitude (table 3.2). At the same time, the hole density $n_h(x)$ rises by two orders of magnitude. Hence, the drift current $j_{\text{drift}}(x) = n_h(x)\,\mu_h(x)\,F(x)$ is constant for both sides of the interface. Using the Einstein relation $D(x) = \mu(x)\,k_B T/q$, the diffusion constant $D_h(x)$ is calculated from the charge carrier mobility $\mu_h(x)$, where the temperature T, the Boltzmann constant k_B and the particle charge q are constants. Hence the diffusion constant also rises by two orders of magnitude at the interface. Since the changes of the diffusion constant

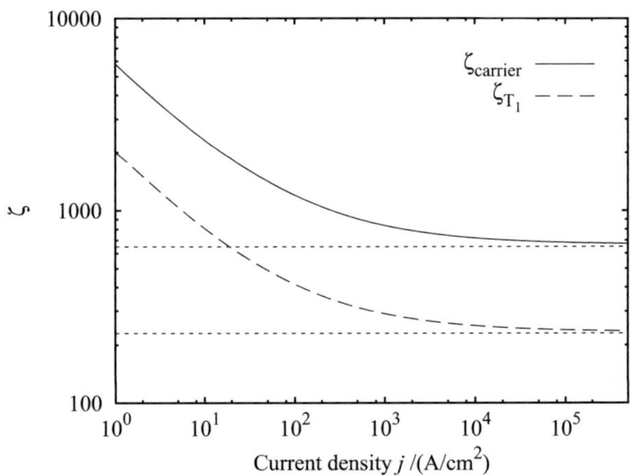

Figure 5.3: ζ_{carrier} (solid line) and $\zeta_{\mathrm{T_1}}$ (dashed line) as a function of current density. For current densities above 10^3 A/cm^2, ζ_{carrier} and $\zeta_{\mathrm{T_1}}$ saturate at a value of 650 and 230, respectively (dotted lines).

and the particle density compensate each other, the diffusion current $j_{\mathrm{diffusion}}(x) = - d/dx\,(D_{\mathrm{h}}(x) \times n_{\mathrm{h}}(x))$ is zero. The same considerations apply to the interface of emission layer and electron transport layer for electrons.

In order to estimate an upper limit for the relative absorption cross sections $\Sigma_{\mathrm{carrier}}$ and $\Sigma_{\mathrm{T_1 T_N}}$, which have been defined in section 3.4.3.3, the particle densities are studied in the absence of lasing. The singlet exciton density is reduced by stimulated emission while the densities of polarons and triplet excitons are unaffected. Hence, in the absence of lasing, lower limits for ζ_{carrier} and $\zeta_{\mathrm{T_1}}$ are calculated. Using equation (3.52) and equation (3.53), upper limits for the relative absorption cross sections $\Sigma_{\mathrm{carrier}}$ and $\Sigma_{\mathrm{T_1 T_N}}$ can be derived.

In figure 5.3, the resulting values for ζ_{carrier} and $\zeta_{\mathrm{T_1}}$ are shown as a function of current density for standard parameters in the absence of lasing. For current densities of less than 10 A/cm^2 both ζ_{carrier} and $\zeta_{\mathrm{T_1}}$ show a linear dependence on the current density in the log-log-plot. For current densities between 10 A/cm^2 and 10^3 A/cm^2 the absolute value of the slope of $\zeta_{\mathrm{carrier}}(j)$ and $\zeta_{\mathrm{T_1}}(j)$ gradually decreases. For current densities above 10^3 A/cm^2 ζ_{carrier} and $\zeta_{\mathrm{T_1}}$ finally saturate at $\zeta_{\mathrm{carrier,sat}} = 650$ and $\zeta_{\mathrm{T_1,sat}} = 230$. This result demonstrates that for the analyzed device, ζ_{carrier} and $\zeta_{\mathrm{T_1}}$ cannot be reduced below the saturation value by increasing the current density. Hence, according to equation (3.52) and equation (3.53), laser operation is only possible if $\Sigma_{\mathrm{carrier}} < 1/650 \approx 1.53 \times 10^{-3}$ and $\Sigma_{\mathrm{T_1 T_N}} < 1/230 \approx 4.34 \times 10^{-3}$ is fulfilled.

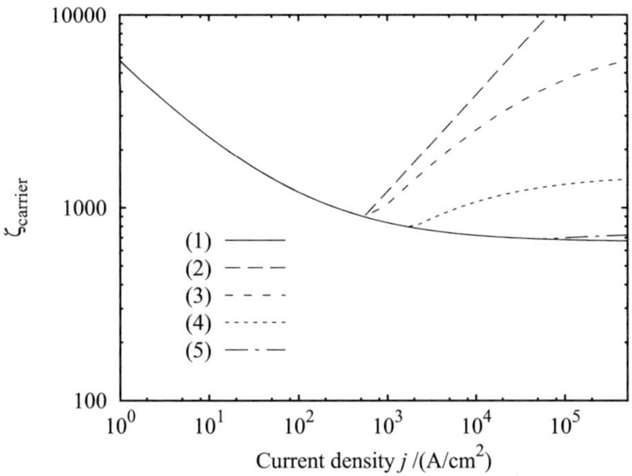

Figure 5.4: ζ_{carrier} as a function of current density for various stimulated and absorption cross sections. For (2) to (5) the cross section for stimulated emission of F8BT ($\sigma_{\text{SE}} - 7 \times 10^{-16}$ cm^2) has been assumed. (1) $\Sigma_{\text{carrier}} = \Sigma_{\text{T}_1\text{T}_\text{N}} = \sigma_{\text{SE}} = 0$, (2) $\Sigma_{\text{carrier}} = \Sigma_{\text{T}_1\text{T}_\text{N}} = 0$, (3) $\Sigma_{\text{carrier}} = \Sigma_{\text{T}_1\text{T}_\text{N}} = 10^{-4}$, (4) $\Sigma_{\text{carrier}} = \Sigma_{\text{T}_1\text{T}_\text{N}} = 5 \times 10^{-4}$, (5) $\Sigma_{\text{carrier}} = \Sigma_{\text{T}_1\text{T}_\text{N}} = 10^{-3}$. For (2) - (5) the curves show a sharp bend at the laser threshold.

The saturation values $\zeta_{\text{carrier,sat}}$ and $\zeta_{\text{T}_1,\text{sat}}$ are decisive factors for the feasibility of laser operation of the considered device. In figure 5.4 ζ_{carrier} is plotted as a function of current density for standard parameters now including the possibility for stimulated emission.

Curve (1) (same curve as solid line in figure 5.3) shows ζ_{carrier} for cross sections of stimulated emission, polaron and triplet-triplet absorption being set to zero. When stimulated emission is set to its standard value ($\sigma_{\text{SE,FSBT}} = 7 \times 10^{-16}$ cm^2), as shown in curve (2), ζ_{carrier} exhibits a sharp bend at the laser threshold current density (j_{thr} = 564 A/cm^2). Above the laser threshold, ζ_{carrier} is increasing linearly in the log-log plot, since the singlet exciton density is constant while the polaron density is increasing with increasing current density.

In curves (3-5), polaron and triplet-triplet absorption are also included, whereby the relative absorption cross sections are set as follows: (3) $\Sigma_{\text{carrier}} = \Sigma_{\text{T}_1\text{T}_N} = 10^{-4}$, (4) $\Sigma_{\text{carrier}} = \Sigma_{\text{T}_1\text{T}_N} = 5 \times 10^{-4}$, (5) $\Sigma_{\text{carrier}} = \Sigma_{\text{T}_1\text{T}_N} = 10^{-3}$. Here, ζ_{carrier} also exhibits a sharp bend at the laser threshold current density. Above the laser threshold, ζ_{carrier} increases and finally saturates. An increasing current density leads to increasing polaron and triplet exciton densities causing an additional attenuation of the laser mode by polaron and triplet-triplet absorption. This additional attenuation is compensated by an increasing singlet exciton density.

For inorganic devices, the effect of polaron absorption is denoted as free carrier absorption. The impact of free carrier absorption on the threshold characteristics has been investigated theoretically and tolerable parameters for laser operation have been predicted [217]. In the presence of free carrier absorption, the existence of a second laser threshold has been derived. In contrast to inorganic lasers, the gain in organic materials is formed by excitons, which are stable at room temperature. Furthermore, lasing occurs at a red shifted spectral position of the vibronic sideband. Thus, the resulting energy scheme is a four level situation. Even under high excitation, most of the molecules are not excited, thus the electron and hole occupancies are much less than unity. Hence, a second threshold is not observed.

5.3 Laser threshold current density

For laser operation the gain in the active material has to outweigh the losses of the laser mode induced by electrode attenuation and carrier as well as triplet exciton related absorption. Here, the impact of induced absorption processes on the threshold current density is analyzed for the TE$_0$-mode laser device design. The threshold current density is determined from the laser characteristics as described in [164]. In figure 5.5 the threshold current density is shown as a function of the relative cross sections for polaron absorption Σ_{carrier} (solid line) and triplet-triplet absorption $\Sigma_{\text{T}_1\text{T}_N}$ (dashed line). In this figure, only one relative absorption coefficient is varied at a time while the other relative absorption coefficient is set to zero. The vertical dotted lines indicate the values $\zeta^{-1}_{\text{carrier,sat}}$ and $\zeta^{-1}_{\text{T}_1,\text{sat}}$ (see section 3.4.3.3).

For relative polaron absorption cross sections $\Sigma_{\text{carrier}} < 5 \times 10^{-5}$, the threshold

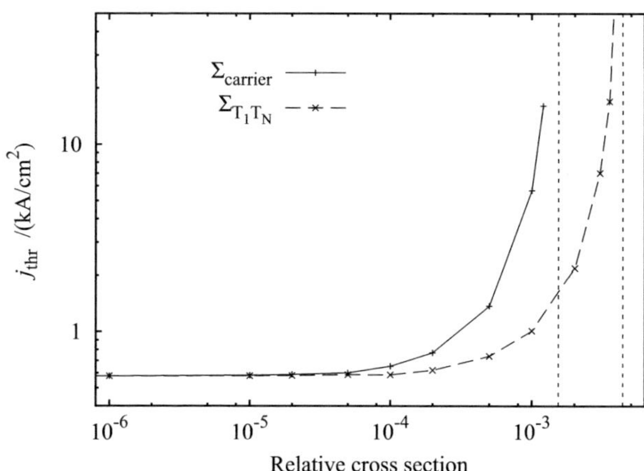

Figure 5.5: Threshold current density j_{thr} as a function of relative polaron absorption cross section Σ_{carrier} (solid line) and as a function of the relative triplet-triplet absorption cross section $\Sigma_{\text{T}_1\text{T}_\text{N}}$ (dashed line). The saturation values $\zeta^{-1}_{\text{carrier,sat}}$ and $\zeta^{-1}_{\text{T}_1\text{T}_\text{N},\text{sat}}$ are indicated by the two vertical dotted lines.

current density j_{thr} is virtually independent of Σ_{carrier}. For values of Σ_{carrier} above 10^{-4}, j_{thr} increases strongly and finally diverges at $\Sigma_{\text{carrier}} = \zeta_{\text{carrier,sat}}^{-1}$. For values of $\Sigma_{\text{carrier}} > \zeta_{\text{carrier,sat}}^{-1}$ lasing is suppressed for all current densities.

The impact of triplet-triplet absorption on the threshold current density j_{thr} is negligible for cross sections $\Sigma_{\text{T}_1\text{T}_\text{N}} < 2 \times 10^{-4}$. j_{thr} is strongly increasing for values of $\Sigma_{\text{T}_1\text{T}_\text{N}} > 2 \times 10^{-4}$ and finally diverges at $\Sigma_{\text{T}_1\text{T}_\text{N}} = \zeta_{\text{T}_1,\text{sat}}^{-1}$. Hence, laser operation is suppressed for cross sections $\Sigma_{\text{T}_1\text{T}_\text{N}} > \zeta_{\text{T}_1,\text{sat}}^{-1}$.

In an actual laser diode structure, both polaron and triplet-triplet absorption are present simultaneously increasing the threshold current density even further. Hence, for laser operation the stronger condition equation (3.51) has to be fulfilled and the relative cross sections for polaron and triplet-triplet absorption, which would still allow laser operation, may be even lower.

5.4 Impact of device geometry

In this sections, the influence of device parameters on ζ is analyzed. ζ depends on the modal intensity profile $I(\text{x})$ and the distribution of polarons and excitons $n(\text{x})$ in the device. $I(\text{x})$ is influenced by the refractive index of the employed organic and electrode materials. The particle densities $n(\text{x})$ are determined by various electronic properties such as mobilities, radiative and non radiative lifetimes as well as annihilation processes. $I(\text{x})$ and $n(\text{x})$ are furthermore influenced by device geometry. Hence the dependence of ζ on device as well as on material properties is very complex. Therefore a parametric study is performed in order to examine the influence of the particular parameters on ζ.

The laser threshold current density strongly depends on the actual relative absorption cross sections, which depend on the wavelength and are still unknown for many organic materials. However, as described in section 3.4.3.3, the knowledge of ζ_{carrier} and ζ_{T_1} as a function of current density can be used to decide whether a specific material is able to show laser operation or not.

Here, the influence of device geometry parameters on ζ is analyzed. The cross sections for stimulated emission, polaron and triplet-triplet absorption are set to zero. ζ_{carrier} and ζ_{T_1} can be used in order to evaluate the capability of a considered laser material and diode structure design for laser operation. From this, maximum cross sections, which would still allow laser operation, can be given. Only one device parameter is varied at a time. With the knowledge of the complete set of actual cross sections, the laser threshold current density can be calculated.

5.4.1 Emission layer thickness

Figure 5.6 shows ζ_{carrier} as a function of current density j for various emission layer thicknesses d_{EML}, whereby the total device thickness is kept constant at 1300 nm. For current densities above 10^4 A/cm^2, ζ_{carrier} saturates at roughly the same value between 600 and 700 for all studied values of d_{EML} ($\zeta_{\text{carrier,sat}} = 650$ for $d_{\text{EML}} = 400$ nm). This

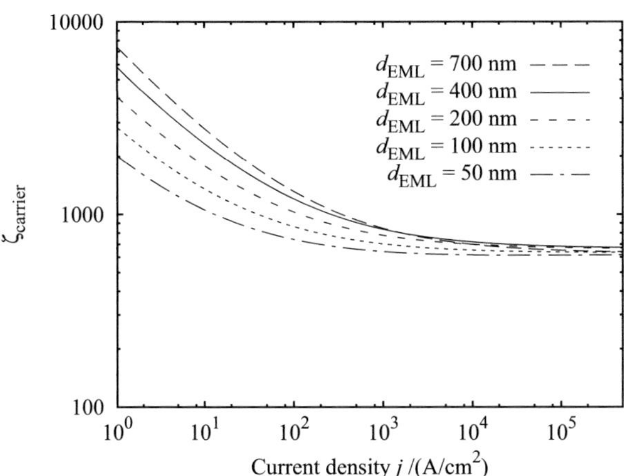

Figure 5.6: ζ_{carrier} as a function of current density for a variation of the emission layer thickness d_{EML}. For current densities below 10^3 A/cm^2, devices with a thin emission layer show a smaller value of ζ_{carrier} than devices with a thicker emission layer. If the current density is larger than 10^4 A/cm^2, ζ_{carrier} saturates at about the same value of 650 for all investigated emission layer thicknesses.

result demonstrates that the impact of polaron absorption cannot be reduced by an optimization of the emission layer thickness.

For current densities below 10^3 A/cm^2, however, thin emission layers provide smaller values of ζ_carrier, which render them favorable with respect to polaron absorption. This result can be understood as follows: In thin emission layers, the particle densities are higher, hence quadratic processes such as Langevin recombination and bimolecular annihilations are much more pronounced. Hence, ζ_carrier approaches its saturation value at lower current densities.

The impact of a variation of d_EML on ζ_{T_1} has also been investigated (not shown). For current densities below 10^2 A/cm^2, ζ_{T_1} is decreasing linearly in the log-log plot for all values of d_EML and thin emission layers also appear to be favorable in this case. This result can again be understood by considering the impact of annihilation processes. In the absence of annihilations, the triplet exciton density is about 10^4 times higher than the singlet exciton density, which is caused by the much longer triplet exciton lifetime (Table 3.2). For the considered annihilation rate coefficients, the triplet exciton density is much more affected by annihilations than the singlet exciton density, hence annihilations reduce ζ_{T_1}. Since the effect of annihilation processes is stronger in thin emission layers, ζ_{T_1} is smaller in this case. For current densities above 10^4 A/cm^2, ζ_{T_1} saturates at roughly the same value for all studied values for d_EML ($\zeta_{T_1} = 230$ for $d_\text{EML} = 400$ nm). Hence, the impact of triplet-triplet absorption cannot be reduced by optimizing the thickness of the emission layer.

5.4.2 Transport layer thickness

In organic DH laser diode structures with thick transport layers, polarons are transported over a longer distance where they can absorb laser light until they form excitons, hence an increased polaron absorption is expected. In figure 5.7, ζ_carrier is shown as a function of current density for a variation of the transport layer thickness. For all current densities, ζ_carrier differs only slightly for laser devices with transport thicknesses between 50 and 1000 nm. This result is remarkable since the distance, where polarons and light can interact in the device, is increased significantly. This result can be understood by analyzing the laser intensity profile I and the spatial distribution of polarons n_e and n_h.

Figure 5.2 shows the normalized modal profile $I(\text{x})$, the polaron densities $n_\text{e}(x)$ and $n_\text{h}(x)$ and the exciton densities $n_{S_1}(x)$ and $n_{T_1}(x)$ at a current density of 10 kA/cm^2. In the transport layers, the intensity of the laser mode is much lower than the intensity in the emission layer. Additionally, the density of polarons is up to two orders of magnitude lower than in the emission layer. This is explained by the much higher mobilities in the transport layers. Hence, polaron absorption mainly takes place in the emission layer and the impact of the transport layer thickness on ζ_carrier is small.

The investigation of ζ_{T_1} demonstrates that ζ_{T_1} is virtually unaffected by d_TL (not shown). In the considered device, both singlet and triplet excitons are assumed to be confined to the emission layer. Hence, the influence of the transport layer thickness d_TL on the triplet exciton density n_{T_1}, as well as on ζ_{T_1}, is marginal.

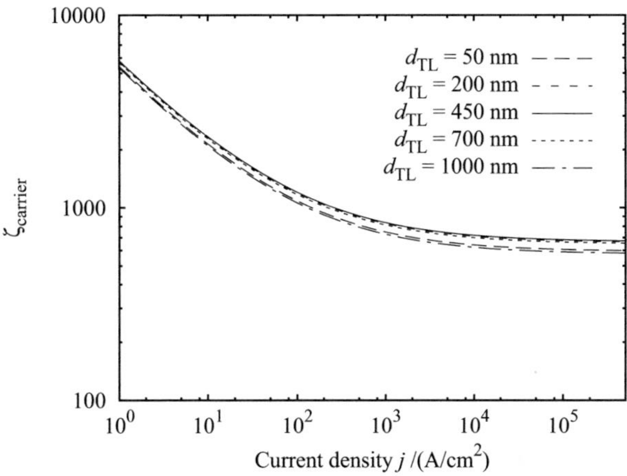

Figure 5.7: ζ_{carrier} as a function of the current density for a variation of d_{TL}. For all current densities, devices with thin transport layers have almost the same value for ζ_{carrier} as devices with a thick transport layer. For current densities above 10 kA/cm^2, ζ_{carrier} saturates at about the same value of 650 for all investigated transport layer thicknesses.

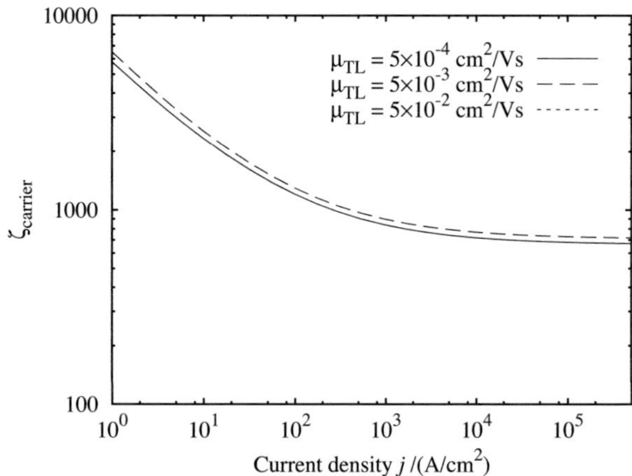

Figure 5.8: ζ_{carrier} as a function of current density for various mobilities in the transport layer. For all current densities the curves are virtually on top of each other. Hence, ζ_{carrier} is not reduced for increasing transport layer mobilities μ_{TL}.

5.5 Impact of material properties

The characteristics of a laser diode structure is not only affected by the device geometry but also by various material parameters. In this section, the effect of material properties on polaron and triplet-triplet absorption, characterized by ζ_{carrier} and ζ_{T_1}, is studied. The influence of charge carrier mobilities of the emission and transport layers on ζ_{carrier} and ζ_{T_1} is analyzed and the role of the exciton lifetimes is discussed. Only one device parameter is varied at a time. With the knowledge of the complete set of actual cross sections, the laser threshold current density can be calculated.

5.5.1 Charge carrier mobilities

Increasing the mobility of organic semiconductors is a promising way to reduce the effect of charge carrier absorption, since the polaron density is decreased throughout the device for a given current density.

5.5.1.1 Transport layer

The impact of the polaron mobility μ_{TL} in the transport layers on ζ_{carrier} is shown in figure 5.8. ζ_{carrier} does not change significantly for all current densities when the mobility of the transport layers is increased by two orders of magnitude from

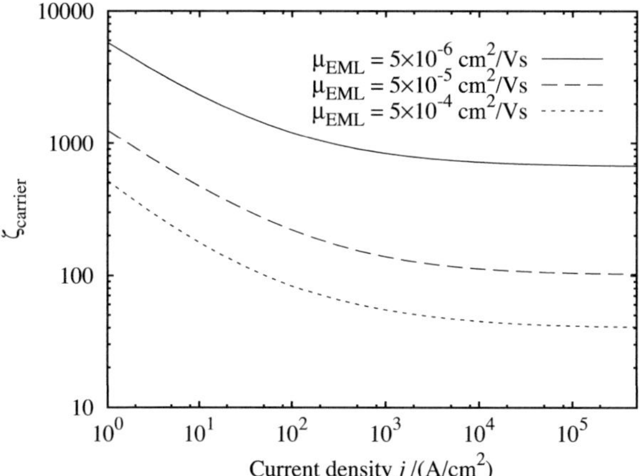

Figure 5.9: ζ_{carrier} as a function of current density for various mobilities in the emission layer, where the electron and hole mobilities are assumed to be equal. ζ_{carrier} is reduced by more than one order of magnitude for μ_{EML} being increased from 5×10^{-6} cm²/Vs to 5×10^{-4} cm²/Vs.

5×10^{-4} cm²/Vs to 5×10^{-2} cm²/Vs. As shown in figure 5.2, the contribution of the transport layers to polaron absorption is marginal for standard parameters, hence ζ_{carrier} cannot be reduced significantly by increasing the mobilities in the transport layers. This is in accord with the findings discussed in section 5.4.2.

5.5.1.2 Emission layer

In figure 5.9, ζ_{carrier} is shown as a function of current density for various mobilities in the emission layer. In this context, the mobilities in the emission layer for electrons and holes are assumed to be equal, hence $\mu_{\mathrm{EML}} = \mu_{\mathrm{EML,h}} = \mu_{\mathrm{EML,e}}$. For increasing mobilities in the emission layer, ζ_{carrier} is strongly reduced for all current densities. When μ_{EML} is increased from 5×10^{-6} cm²/Vs to 5×10^{-5} cm²/Vs, ζ_{carrier} is reduced by a factor of 5 from 2280 to 460 at a current density of 10 A/cm². Additionally, the saturation value $\zeta_{\mathrm{carrier,sat}}$ is also reduced at high current densities: at 10^4 A/cm² ζ_{carrier} is reduced by a factor of $\approx$ 7 from 710 to 110. By further increasing μ_{EML} to 5×10^{-4} cm²/Vs, the saturation value $\zeta_{\mathrm{carrier,sat}}$ is only reduced by a factor of 2.5. In this case, the mobilities of the transport layers and the emission layer are equal. Thus, the polaron densities are of the same order of magnitude, hence the relative contribution of the transport layers to the total polaron absorption has increased. This result shows that the mobilities in the emission layer and in the transport layers have to be increased simultaneously in order to reduce $\zeta_{\mathrm{carrier,sat}}$ further.

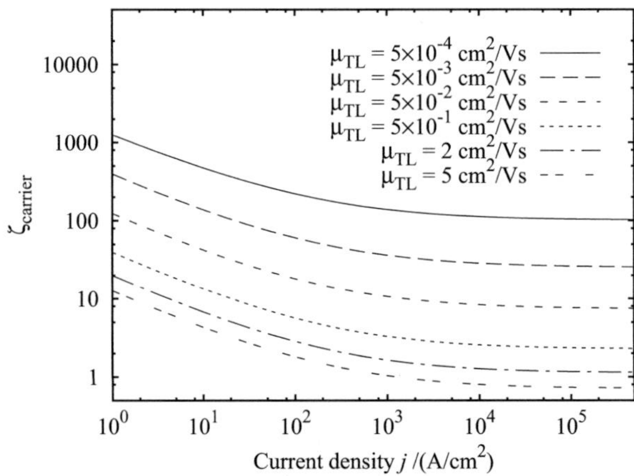

Figure 5.10: ζ_{carrier} is shown as a function of the current density for various mobilities in the emission and in the transport layer, whereby $\mu_{\mathrm{TL}} = 10{\times}\mu_{\mathrm{EML}}$. By increasing the mobilities to $\mu_{\mathrm{TL}} = 2$ cm^2/Vs and $\mu_{\mathrm{EML}} = 0.2$ cm^2/Vs, ζ_{carrier} is reduced to 1 for current densities above 10^3 A/cm^2.

5.5.1.3 Emission and transport layer

In the last two paragraphs the impact of an improved mobility in the emission and in the transport layers has been studied, whereby only one mobility has been varied at a time. In this paragraph, we study the effect of charge carrier mobilities on ζ_{carrier} when they are increased in all organic layers simultaneously. In contrast to the standard parameters, we now assume that the charge carrier mobilities in the transport layers is one order of magnitude higher than the mobility in the emission layer. By increasing the mobilities, ζ_{carrier} can be reduced significantly for all current densities as shown in figure 5.10. By increasing μ_{TL} to 2 cm^2/Vs ($\mu_{\mathrm{EML}} = 0.2$ cm^2/Vs), the saturation value of ζ_{carrier} can be reduced to 1. In this case, laser operation may be possible even if the cross sections for stimulated emission and charge carrier absorption are of the same order of magnitude.

For most organic materials, typical hole mobilities are in the range from 10^{-6} cm^2/Vs to 10^{-4} cm^2/Vs [194, 196, 220]. Typical electron mobilities are in the range from 10^{-8} cm^2/Vs to 10^{-5} cm^2/Vs [198,221]. Recently, hole mobilities of up to 10^{-2} cm^2/Vs [157,196] and electron mobilities of up to 10^{-3} cm^2/Vs [197] have been reported.

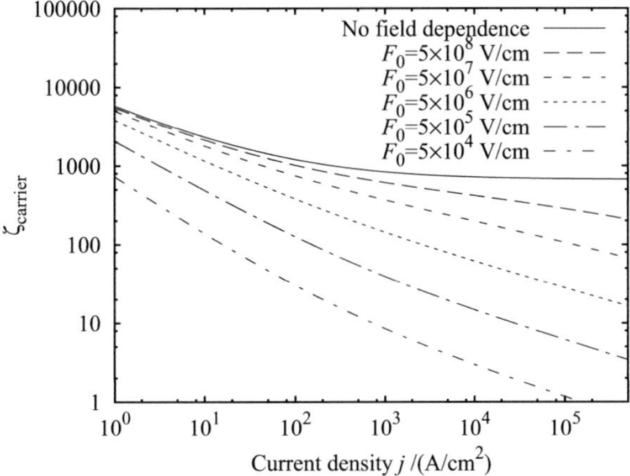

Figure 5.11: ζ_{carrier} as a function of the current density for various field dependence coefficients F_0 of the mobility in the emission and the transport layer. In the case of field dependent mobilities, ζ_{carrier} does not saturate at high current densities.

5.5.1.4 Field dependent mobilities

Charge carrier mobilities in organic materials usually exhibit a field dependency [197, 222, 223]. For fields of about 0.5 MV/cm, many organic materials show a good agreement with a Poole-Frenkel law

$$\mu_{\text{e,h}} = \mu_{0;\text{e,h}} \exp\left(\sqrt{F/F_0}\right).$$

Here, μ_0 describes the zero-field mobility of charge carriers and F_0 is a parameter which describes the field dependence of the mobility. Typical materials exhibit a field dependence from $F_0 = 5\times10^4$ V/cm to $F_0 = 5\times10^6$ V/cm [223]. However, high-mobility materials show only a very small field dependence [195, 197]. Since high mobility materials are preferred in organic laser diodes the field dependence had been set to zero. Assuming that the employed high-mobility materials still exhibit field dependent mobilities at fields above 1 MV/cm and that the materials of the emission layer and the transport layer exhibit the same field dependencies, the influence on ζ_{carrier} is plotted in figure 5.11, where ζ_{carrier} is strongly reduced for current densities above 10 A/cm^2. For increasing current densities, the fields and the mobilities are increasing causing ζ_{carrier} not to saturate. Hence typical field dependencies of the mobilities lead to a reduction of ζ_{carrier}.

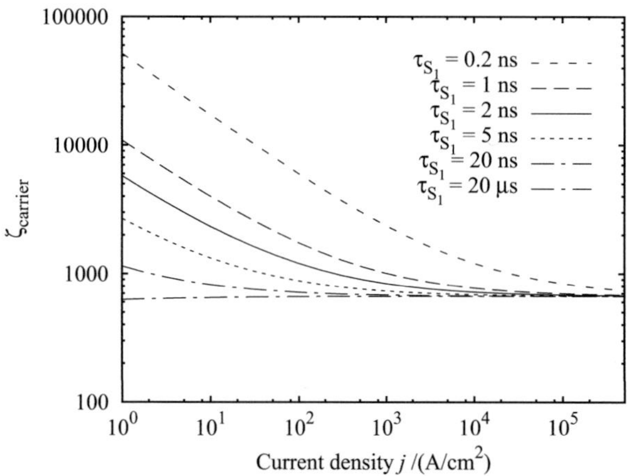

Figure 5.12: ζ_{carrier} as a function of current density for various singlet exciton lifetimes τ_{S_1}. By increasing the singlet exciton lifetime, the saturation value of ζ_{carrier} is reached already at lower current densities.

5.5.2 Variation of exciton lifetimes

The total lifetime of the excitons is determined by their radiative lifetime and mono-molecular as well as bimolecular nonradiative lifetimes.

In this section the impact of the (total monomolecular) lifetimes τ_{S_1} and τ_{T_1} of singlet and triplet excitons on the dimensionless quantities ζ_{carrier} and ζ_{T_1} is studied. In the considered device with standard parameters, the radiative lifetime of triplet excitons exceeds the singlet exciton lifetime by four orders of magnitude, causing n_{T_1} to exceed n_{S_1} by more than two orders of magnitude at a current density of 10^4 A/cm^2 (see figure 5.2).

5.5.2.1 Singlet exciton lifetime

In figure 5.12, ζ_{carrier} is plotted as a function of current density for a variation of the effective exciton lifetime τ_{S_1}, which includes radiative and nonradiative decays. If τ_{S_1} is decreased from its standard value of 2 ns (solid line) to 0.2 ns, ζ_{carrier} is increasing from 2200 to about 16000 ($j = 10$ A/cm^2). If τ_{S_1} is increased to 20 ns, ζ_{carrier} is decreasing to about 800 ($j = 10$ A/cm^2). For all considered lifetimes ζ_{carrier} is decreasing and finally saturating at $\zeta_{\text{carrier}} = 650$, whereby ζ_{carrier} reaches its saturation value at lower current densities for higher singlet exciton lifetimes. Hence, by reducing nonradiative decay channels and therefore increasing the effective monomolecular lifetime, the saturation value for ζ_{carrier} is reached at lower current densities. However, the saturation

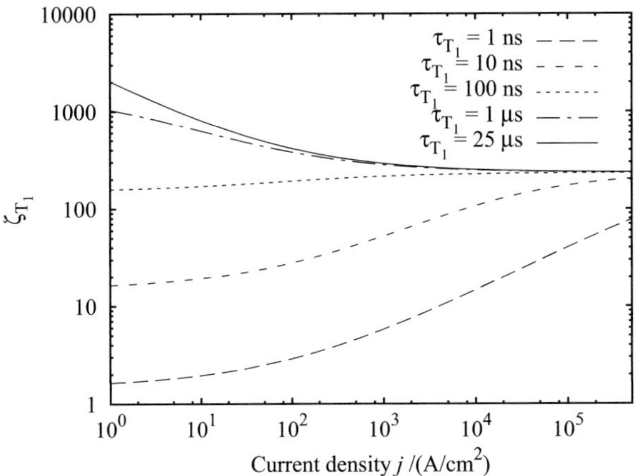

Figure 5.13: ζ_{T_1} as a function of current density for various triplet exciton lifetimes τ_{T_1}. By reducing the monomolecular triplet exciton lifetime τ_{T_1} to 1 ns, ζ_{T_1} can be reduced to almost 1.

value for ζ_{carrier} is not improved. This behavior can be understood by considering that the total lifetime is determined by monomolecular and bimolecular decay channels. At current densities around 1 A/cm^2 the influence of bimolecular annihilations is low, hence the exciton density is mainly determined by the monomolecular lifetime. For increasing singlet exciton lifetimes, the singlet exciton density is increasing, hence ζ_{carrier} is decreasing. At high current densities, bimolecular annihilation dominates, hence ζ_{carrier} is determined by annihilations.

As the lifetime τ_{S_1} is varied, ζ_{T_1} is also influenced (not shown). At low current densities the singlet exciton density is increased for increasing lifetimes τ_{S_1}, hence ζ_{T_1} is decreased. For high current densities, ζ_{T_1} is again saturating at the same value for the considered singlet-exciton lifetimes.

5.5.2.2 Triplet exciton lifetime

For standard parameters, singlet excitons have a much shorter lifetime ($\tau_{S_1} = 2$ ns) than triplet excitons ($\tau_{T_1} = 25$ µs). Since singlet excitons are annihilated by triplet excitons and the laser mode is attenuated by triplet-triplet absorption, it would be desirable to reduce the triplet exciton density. This can be done, for example, by triplet quenchers or phosphorescent emitters.

The role of the monomolecular triplet lifetime is studied in figure 5.13, where ζ_{T_1} is plotted as a function of current density j for various triplet lifetimes τ_{T_1}. At a current density of 1 A/cm^2, ζ_{T_1} is greatly reduced from about 2000 to 150, if τ_{T_1} is decreased from 25 µs to 100 ns. For current densities from 1 to 10 kA/cm^2, ζ_{T_1} is only reduced by less than a factor of 1.5, if τ_{T_1} is decreased from 25 µs to 100 ns. Hence, for current densities above 1 kA/cm^2, triplet exciton quenchers with a lifetime of less than 10 ns are required in order to reduce ζ_{T_1} significantly. By decreasing τ_{T_1} to 1 ns, ζ_{T_1} is reduced to almost 1 for current densities below 10^2 A/cm^2. Since at high current densities ζ_{T_1} is influenced mainly by annihilation processes, the saturation value for ζ_{T_1} is not affected by τ_{T_1}. In OLEDs, phosphorescent emitters are used in order to increase the quantum efficiency [224]. In organic lasers, these triplet emitters may be useful in order to reduce the density of triplet excitons. However, typical triplet emitters exhibit lifetimes in the range from 1 μs to 1 ms. In vapor phase dye lasers, the maximum operation time was increased by using trans-stylbene with a lifetime of about 5 ns as a triplet quencher [225].

By reducing the triplet exciton density, the effect of singlet-triplet annihilation is decreased, so the singlet exciton density rises. On the other side, triplet-triplet annihilation, where singlet excitons are generated, is also reduced. Hence, depending on the annihilation rate coefficients, ζ_{T_1} is also altered. For standard parameters, both effects compensate each other, hence the impact on ζ_{T_1} is only marginal (not shown).

5.6 Pulsed operation

In this section, we show that induced absorption losses by triplet excitons can be decreased considerably by pulsed operation of organic laser diodes. By analyzing the switch-on characteristics of an organic laser diode at a current density of 1 kA/cm^2, we elucidate that the separation of singlet and triplet excitons in the time domain is feasible. A sufficiently high singlet exciton density can be achieved before the density of triplet excitons approaches its steady-state value. Hence, pulsed operation can be employed in order to reduce triplet exciton absorption in organic laser diodes.

5.6.1 Transient absorption parameters

The transient impact of induced absorption processes is described by the absorption parameters ζ_{carrier} and ζ_{T_1} describing the relative amount of polarons and triplet excitons (weighted with the modal intensity profile) compared to singlet excitons. For polaron absorption, the absorption parameter is

$$\zeta_{\text{carrier}}(t) = \frac{\int_0^d I(x)\left(n_e(x,t) + n_h(x,t)\right)dx}{\int_0^d I(x) \cdot n_{S_1}(x,t)\,dx}. \tag{5.1}$$

The corresponding absorption parameter for triplet absorption is

$$\zeta_{T_1}(t) = \frac{\int_0^d I(xt)\,n_{T_1}(x,t)\,dx}{\int_0^d I(x)\,n_{S_1}(x,t)\,dx}. \tag{5.2}$$

For simplicity, we assume that the absorption cross sections for holes and electrons are equal. As deduced in equations 3.52 and 3.53, the necessary conditions $\zeta_{\text{carrier}} < \Sigma^{-1}_{\text{carrier}}$ and $\zeta_{\text{T}_1} < \Sigma^{-1}_{\text{T}_1\text{T}_N}$ have to be fulfilled for laser operation. Since the relative absorption cross sections Σ_{carrier} and $\Sigma_{\text{T}_1\text{T}_N}$ are constants, the absorption parameters ζ_{carrier} and ζ_{T_1} can be used to compare different laser devices and modes of operation concerning their polaron and triplet absorption characteristics.

In an organic laser diode, the key features of the switch-on process are the time dependent modal gain and the absorption parameters. The values of the modal gain and absorption parameters under CW operation are $g_{\text{SE,CW}}$, $\zeta_{\text{T}_1,\text{CW}}$ and $\zeta_{\text{carrier,CW}}$. Characteristic times $t_{g,0.9}$, $t_{\text{T}_1,0.5}$ and $t_{\text{carrier},2}$ are defined as the moments, when the modal gain is $g_{\text{SE}} = 0.9 \times g_{\text{SE,CW}}$, the triplet absorption parameter is $\zeta_{\text{T}_1} = 0.5 \times \zeta_{\text{T}_1,\text{CW}}$ and the polaron absorption parameter is $\zeta_{\text{carrier}} = 2.0 \times \zeta_{\text{carrier,CW}}$. The impact of triplet absorption can be reduced by pulsed operation if the following conditions are fulfilled

$$t_{g,0.9} < t_{\text{T}_1,0.5} \tag{5.3}$$

$$t_{\text{carrier},2} < t_{\text{T}_1,0.5}. \tag{5.4}$$

In this case, the modal gain g_{SE} and the polaron absorption parameter ζ_{carrier} are close to their values for CW operation, while the triplet absorption parameter ζ_{T_1} is still low.

5.6.2 Switch-on characteristics of an organic laser diode

In figure 5.14, the modal gain g_{SE} and the absorption parameters ζ_{carrier} and ζ_{T_1} are plotted as a function of time, when a current step is applied ($j = 1$ kA/cm^2, $t_{\text{rise}} = 5$ ns). During the first 40 ns, the exciton density is virtually zero, since the charge carriers first have to travel through the transport layers before they can recombine in the emission layer. Hence, ζ_{carrier} is virtually infinity. After 50 ns, excitons are formed by Langevin recombination in the emission layer and the modal gain g_{SE} rises while the polaron absorption parameter ζ_{carrier} drops to its saturation value. Triplet excitons accumulate slowly causing ζ_{T_1} to rise. For the considered device, the following characteristic times are determined: $t_{\text{carrier},2} = 61$ ns, $t_{g,0.9} = 72$ ns and $t_{\text{T}_1,0.5} = 94$ ns, hence the conditions (5.3) and (5.4) are fulfilled and the impact of triplet absorption can be reduced for the considered device. At $t_{g,0.9} = 72$ ns, when the modal gain reaches 90% of its steady state value, the absorption parameters are $\zeta_{\text{carrier}} = 1.17 \times \zeta_{\text{carrier,CW}}$ and $\zeta_{\text{T}_1} = 0.21 \times \zeta_{\text{T}_1,\text{CW}}$. Thus, the impact of triplet exciton absorption can be reduced by a factor of 5 for the standard device, while the modal gain and the polaron absorption are already close to their CW operation values. This result is notable, since it shows that a separation of the singlet and the triplet exciton densities is feasible in the time domain.

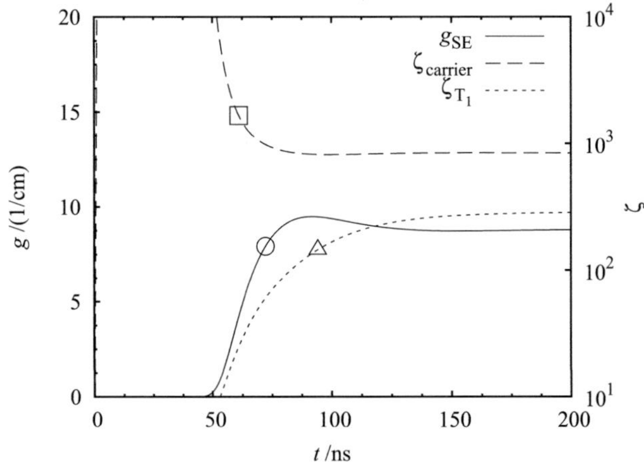

Figure 5.14: Modal gain g_{SE}, polaron absorption parameter ζ_{carrier} and triplet absorption parameter $\zeta_{\mathrm{T_1}}$ as a function of time. The switch-on process is specified by the characteristic times $t_{g,0.9}$ (circle), $t_{\mathrm{carrier},2}$ (square) and $t_{\mathrm{T_1},0.5}$ (triangle).

5.6.3 Charge carrier mobility

In order to deduce design rules for a further reduction of triplet absorption, a parametric study is performed which examines the influence of the device geometry and material properties on the pulsed absorption parameters $\zeta_{\mathrm{T_1,pulsed}} = \zeta_{\mathrm{T_1}}(t_{g,0.9})$ and $\zeta_{\mathrm{carrier,pulsed}} = \zeta_{\mathrm{carrier}}(t_{g,0.9})$.

The triplet absorption parameter is not significantly influenced by the thicknesses of the emission and the transport layers (not shown). In fact, all characteristic times are strongly decreased for thinner devices. Since all characteristic times are affected simultaneously, the singlet and triplet exciton densities cannot be further separated in the time domain and the triplet absorption parameter $\zeta_{\mathrm{T_1}}$ is not further reduced.

Figure 5.15 shows the impact of the charge carrier mobility μ_{EML} in the emission layer on $\zeta_{\mathrm{T_1,CW}}$ and $\zeta_{\mathrm{T_1,pulsed}}$. When μ_{EML} is increased, $\zeta_{\mathrm{T_1,pulsed}}$ is strongly reduced while $\zeta_{\mathrm{T_1,CW}}$ remains almost constant. This behavior can be understood as follows: For higher mobilities in the emission layer, Langevin recombination is strongly enhanced. Hence the rise time of the modal gain g_{SE} is much shorter, and the triplet excitons have less time to accumulate in the device. As a result the triplet absorption parameter $\zeta_{\mathrm{T_1}}$ decreases. For $\mu_{\mathrm{EML}} = 5 \times 10^{-4}\,\mathrm{cm^2/Vs}$, the triplet absorption parameter $\zeta_{\mathrm{T_1}}$ can be reduced by a factor of 60 from $\zeta_{\mathrm{T_1,CW}} = 316$ to $\zeta_{\mathrm{T_1,pulsed}} = 5.3$. In conclusion, for pulsed operation higher mobilities in the emission layer are favorable in terms of triplet absorption, while the effect of μ_{EML} on $\zeta_{\mathrm{T_1}}$ is negligible for CW

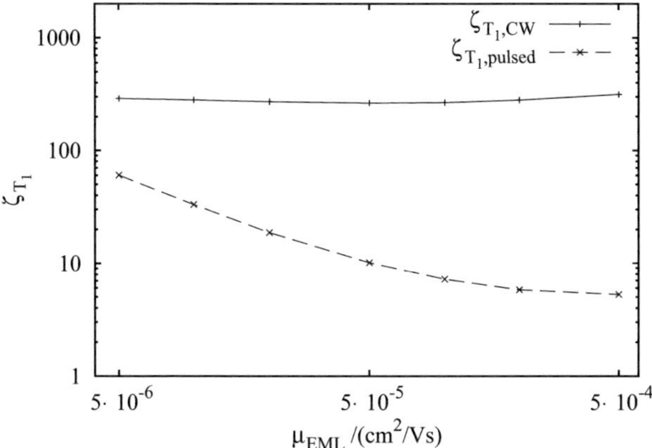

Figure 5.15: The triplet absorption parameter for CW operation $\zeta_{\mathrm{T_1,CW}}$ and the triplet absorption parameter for pulsed operation $\zeta_{\mathrm{T_1,pulsed}} = \zeta_{\mathrm{T_1}}(t_{g,0.9})$ as a function of the charge carrier mobility in the emission layer μ_{EML}. For mobilities of $5 \times 10^{-4}\,\mathrm{cm^2/Vs}$, $\zeta_{\mathrm{T_1}}$ can be reduced to $1/60$ of its CW value.

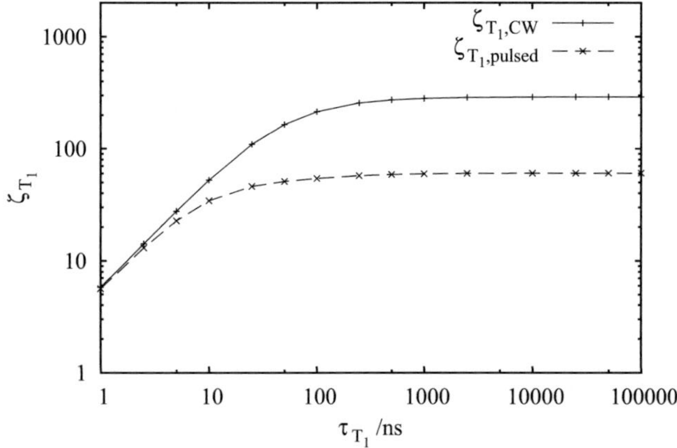

Figure 5.16: The triplet absorption parameters for CW and pulsed operation $\zeta_{T_1,CW}$ and $\zeta_{T_1,\text{pulsed}}$ as a function of the triplet exciton lifetime τ_{T_1}.

operation.

5.6.4 Triplet quenchers

Alternatively to pulsed operation, the use of triplet quenchers is a promising approach to reduce the impact of triplet absorption in an organic laser diode. For CW operation, a triplet quencher with a lifetime of less than 1 ns is necessary in order to reduce $\zeta_{T_1,CW}$ to unity [177]. Here we address the benefit of pulsed operation in the presence of triplet quenchers. In figure 5.16, $\zeta_{T_1,CW}$ and $\zeta_{T_1,\text{pulsed}}$ are plotted as a function of the triplet exciton lifetime. If $\tau_{T_1} > 100$ ns, ζ_{T_1} is reduced significantly by pulsed operation. If τ_{T_1} is decreased below 100 ns, both $\zeta_{T_1,CW}$ and $\zeta_{T_1,\text{pulsed}}$ are reduced, whereby the difference between $\zeta_{T_1,CW}$ and $\zeta_{T_1,\text{pulsed}}$ gradually vanishes. For $\tau_{T_1} < 10$ ns, $\zeta_{T_1,CW}$ and $\zeta_{T_1,\text{pulsed}}$ are virtually equal. In this case, the impact of triplet absorption cannot be further reduced by pulsed operation.

5.6.5 Reverse current pulses

In this section, material properties are employed which are geared to the best available organic semiconductors. We have set the charge carrier mobility to 5×10^{-3} cm^2/Vs [196,197] and the exciton binding energies to $E_{S_1} = 1.4$ eV and $E_{T_1} = 1.6$ eV [109–112], respectively. Charge carrier mobilities and exciton binding energies of this magnitude have already been demonstrated for organic semiconductors. In addition, thinner

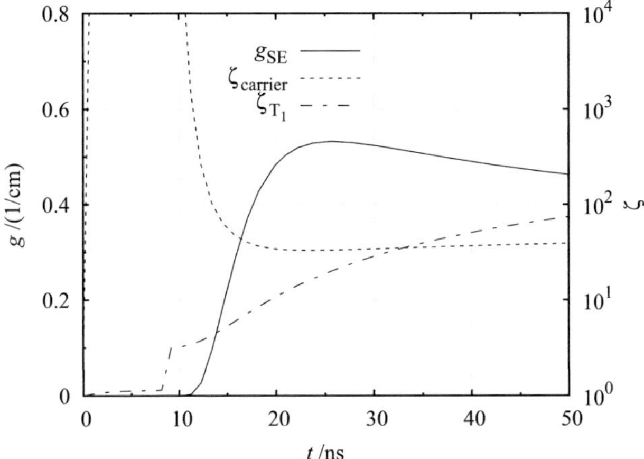

Figure 5.17: Transient characteristics of g_{SE}, ζ_{carrier} and $\zeta_{\mathrm{T_1}}$, when a current step of 30 A/cm^2 is applied.

organic layers ($d_{\mathrm{ETL}} = d_{\mathrm{HTL}} = 150$ nm, $d_{\mathrm{EML}} = 20$ nm) are employed in order to reduce the transit times of the charge carriers.

In figure 5.17, the switch-on characteristics of g_{SE}, ζ_{carrier} and $\zeta_{\mathrm{T_1}}$ are shown as a function of time when a current step is applied ($j = 30$ A/cm^2, $t_{\mathrm{rise}} = 5$ ns). After t = 0 ns, charge carriers are injected but neither singlet nor triplet excitons are generated during the first 10 ns. Electrons and holes first have to drift through the transport layers before excitons can be generated by Langevin recombination in the emission layer. Therefore, ζ_{carrier} is virtually infinite during the first 10 ns. At t = 11 ns, excitons are formed and ζ_{carrier} approaches its steady-state value. g_{SE} reaches its maximum value at $t = 25$ ns, while $\zeta_{\mathrm{T_1}}$ is still almost one magnitude smaller than its steady state value. This shows that the impact of triplet absorption can be reduced by pulsed operation.

In order to reduce the impact of polaron absorption, the first pulse in forward direction (same as above) is followed by a second pulse in reverse direction ($j = -240$ A/cm^2, $t_{\mathrm{rise}} = 2$ ns). In figure 5.17, the transient characteristics of g_{SE}, ζ_{carrier} and $\zeta_{\mathrm{T_1}}$ are shown. At t = 31 ns, polarons are extracted from the device. When the current density of the reverse biased pulse is chosen properly, the time for polaron extraction is smaller than the lifetime of the singlet excitons. At this point, singlet exciton binding energies of less than 0.5 eV are unfavorable because of increased field quenching during the reverse biased pulse. At t = 35 ns, ζ_{carrier} drops below 1 while g_{SE} is 0.3 /cm and $\zeta_{\mathrm{T_1}}$ is still small. In this case, optical gain could be achieved if $\sigma_{\mathrm{T_1T_N}} < 10^{-3} \cdot \sigma_{\mathrm{SE}}$ [11] and $\sigma_{\mathrm{carrier}} \approx \sigma_{\mathrm{SE}}$ [226].

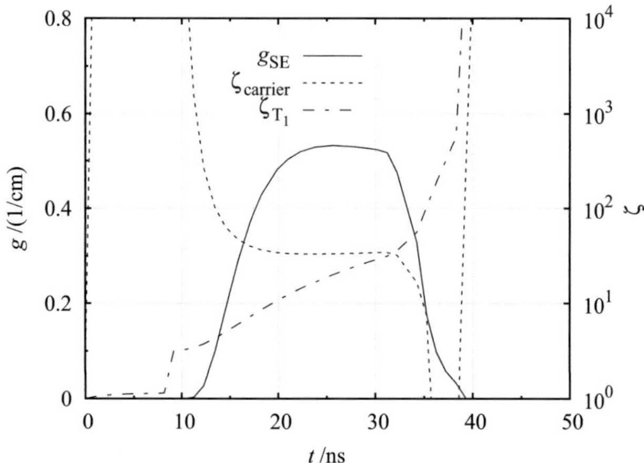

Figure 5.18: g_{SE}, ζ_{carrier} and $\zeta_{\mathrm{T_1}}$ as a function of time, when a current pulse of 30 A/cm^2 is applied, which is followed by a reverse current pulse of -240 A/cm^2.

5.6.6 Summary

In summary, we have studied the properties of polaron and triplet absorption in organic double-heterostructure laser diodes under pulsed excitation using numerical simulations. At a current density of 1 kA/cm^2, singlet and triplet excitons can be separated in the time domain and the impact of triplet absorption can be reduced by a factor of 5. By increasing the charge carrier mobility in the emission layer, the absorption by triplet excitons can be further decreased by a factor of 60. The impact of triplet absorption can also be decreased by reducing the effective triplet exciton lifetime using triplet quenchers. If the effective lifetime of triplet excitons is below 10 ns, triplet absorption is not further reduced by pulsed operation. Reverse biased pulses might be able to separate singlet excitons from polarons in the time domain. However, this approach is technologically challenging.

5.7 Chapter summary

We have investigated the impact of charge carrier and triplet exciton absorption on the threshold current density of an organic DH laser diode structure using numerical simulations. In our model, standard parameters, as known from materials used for highly efficient organic light emitting diodes, have been employed. Absorption parameters $\zeta_{carrier}$ and ζ_{T_1} have been introduced which quantify the effect of polaron and excited-

state absorption in the device. For the polaron and the triplet-triplet absorption cross sections, upper limits of $\sigma_{\mathrm{carrier}} = 1.53 \times 10^{-3} \times \sigma_{SE}$ and $\sigma_{T_1 T_N} = 4.34 \times 10^{-3} \times \sigma_{SE}$ have been deduced.

If annihilations are included in the model, ζ eventually saturates, which implies that polaron and triplet-triplet absorption can prevent electrically pumped devices from lasing for all current densities. The impact of various device and material parameters on ζ has been investigated. It was shown that ζ does not strongly depend on the device geometry.

For the studied devices, an increased charge carrier mobility in the transport layers does not reduce polaron absorption significantly. If the mobilities in the emission layer and in the transport layers are increased simultaneously, the effect of polaron absorption will be reduced. Minimum mobilities of about 2 cm^2/Vs in the transport layers and about 0.2 cm^2/Vs in the emission layer are necessary if the cross sections for polaron absorption and for stimulated emission are within the same order of magnitude.

The impact of the lifetimes of singlet and triplet excitons on ζ has also been analyzed. By increasing the singlet exciton lifetime, the saturation value of ζ is reached at lower current densities. If the triplet exciton lifetime is decreased to about 1 ns, ζ_{T_1} can be reduced to almost 1 for current densities below 100 A/cm^2. In this case, laser operation may be possible for the cross sections of stimulated emission and triplet-triplet absorption being of the same order of magnitude.

The effect of polaron and triplet-triplet absorption can also be reduced by separating singlet excitons from triplet excitons and polarons in the time domain by pulsed operation and reverse current pulses. This approach has also been investigated in this chapter. The threshold current density is also influenced by electric field-induced quenching, which will be included in the next chapter.

6 Impact of field-induced exciton dissociation

Due to low charge carrier mobilities in organic materials, electric fields of more than 10^6 V/cm are required in order to achieve current densities of the order of 1 kA/cm^2. Thus, an important additional process also increasing the laser threshold is field-induced dissociation of excitons being the species which is responsible for optical gain in organic semiconductors. In this section, organic laser diode structures are discussed, which support the propagation of the TE$_2$-mode [167]. In comparison with the TE$_0$-concept, which has been discussed in chapter 4 and in chapter 5, the TE$_2$ structure exhibits lower operation voltages.

This chapter is structured as follows: In section 6.1, the device geometry of the TE$_2$-mode laser structure is discussed. The threshold current density of the TE$_2$-mode device is calculated in section 6.2. The impact of bimolecular annihilation processes is studied in section 6.2.1. The importance of low waveguide attenuation is clarified in section 6.2.2. The impact of field quenching is studied in section 6.2.3. The impact of the device geometry and various material parameters on the exciton density are studied in section 6.3 and in section 6.4. Implications on the impact of polaron and triplet absorption are given in section 6.5. Guest-host systems, which are less affected by field-induced exciton dissociation, are analyzed in section 6.6. The results of this chapter are summarized in section 6.7.

6.1 Device geometry

In this section, the investigated laser diode structure (figure 6.1) is described [167]. The considered organic double heterostructure (DH) device is formed by transport layers for holes (HTL) and electrons (ETL) and an emission layer (EML). The layer thickness is 150 nm for the transport layers and 20 nm for the emission layer. Excitons are assumed to be confined to the emission layer and electrons and holes are blocked by the opposite transport layer, respectively. Transparent conductive oxides are used as electrodes exhibiting attenuation coefficients of 150 cm^{-1} (anode) and 750 cm^{-1} (cathode) [168].

In a preceding study, we have demonstrated by numerical simulations that a low attenuation coefficient of the waveguide is of crucial importance for laser operation [164]. This has also been demonstrated experimentally for optically pumped devices with highly conductive electrodes [50, 89, 165, 199]. In order to decrease waveguide attenuation, the filling factor of the laser mode within the electrodes has to be small.

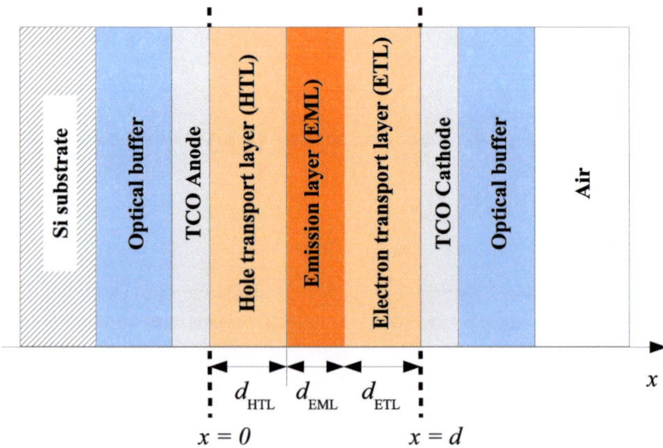

Figure 6.1: The simulated double heterostructure laser diode with hole transport layer (HTL), emission layer (EML) and electron transport layer (ETL).

Thus, optical buffer layers are employed enabling the propagation of the $\mathrm{TE_2}$-mode. Since the electrodes are within the minima of the modal intensity profile, waveguide losses of about 4.3 $\mathrm{cm^{-1}}$ can be achieved [167]. The modal intensity profile of the $\mathrm{TE_2}$-mode is shown in figure 6.2. For the emission layer, the confinement factor is 1.9 %.

6.2 Threshold current density

In this section, the threshold current density is calculated in the presence of various loss processes. Section 6.2.1 shows, that laser operation can still be achieved in the presence of bimolecular annihilation processes. The necessity of low waveguide attenuation is elucidated in section 6.2.2. In section 6.2.3, both, bimolecular annihilation processes and field quenching are considered. In this case laser operation is not achieved.

6.2.1 Bimolecular annihilation processes

In figure 6.3, the threshold current density j_{thr} is shown as a function of the emission layer thickness d_{EML}, when annihilation processes are included and in their absence. The thickness of the emission layer is varied while the total device thickness is kept constant at 320 nm. In the absence of loss processes, laser operation is achieved at a current density of $j_{\mathrm{thr}} = 206$ A$/\mathrm{cm^2}$ ($U_{\mathrm{thr}} = 158$ V) for an emission layer thickness of $d_{\mathrm{EML}} = 20$ nm (solid curve). In this case, j_{thr} is virtually independent of the

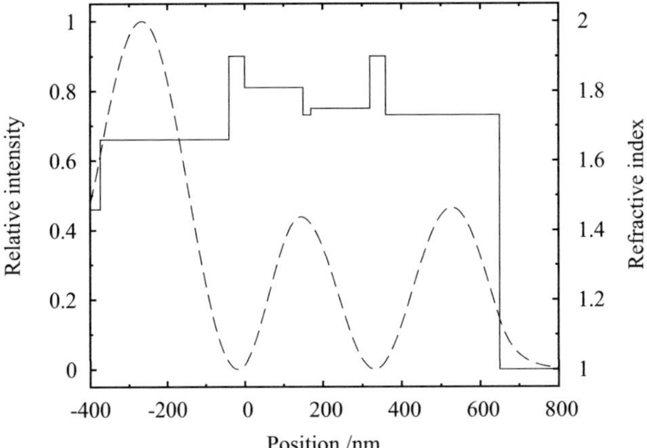

Figure 6.2: Modal intensity profile and refractive index of the employed materials as a function of position. The employed materials, refractive indices and layer thicknesses are specified in table 3.1.

emission layer thickness. When annihilation processes are included, j_{thr} strongly increases (dashed curve): At $d_{\text{EML}} = 20$ nm, the laser threshold is 8.5 kA/cm^2 ($U_{\text{thr}} = 868$ V). Including annihilations, the dependence of j_{thr} on the emission layer thickness is pronounced. When d_{EML} is increased to 100 nm, j_{thr} decreases to 2.3 kA/cm^2 ($U_{\text{thr}} = 860$ V). For $d_{\text{EML}} = 5$ nm, j_{thr} increases to 30 kA/cm^2 ($U_{\text{thr}} = 1.47$ kV). This result demonstrates that thin emission layers are unfavorable regarding annihilation processes. For emission layer thicknesses of more than 20 nm, excitons are not generated homogeneously over the emission layer due to unequal mobilities for electrons and holes [227]. Therefore, all simulations are carried out for an emission layer thickness of 20 nm.

6.2.2 Waveguide attenuation

We have demonstrated in [164] that a low attenuation coefficient of the waveguide is inevitable for laser operation. This has also been shown experimentally for optically pumped devices with highly conductive electrodes [165]. In our model, the organic layers are assumed to be transparent while the TCO electrodes exhibit an attenuation coefficient of 150 cm^{-1} for the bottom layer and 750 cm^{-1} for the top layer [168]. While transfer-matrix simulations predict a waveguide attenuation coefficient of about 1 cm^{-1}, experimental results provide attenuation coefficients of 4.3 cm^{-1} [167]. In our calculations, the waveguide properties have been adapted to the experimental results.

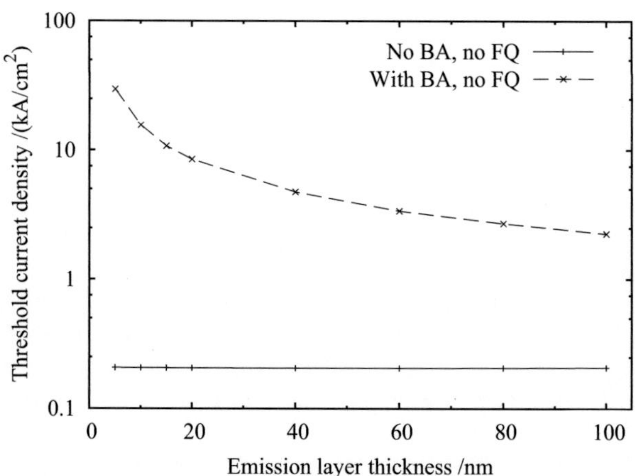

Figure 6.3: Laser threshold current density as a function of the emission layer thickness, when annihilation processes are included and in their absence. In the presence of bimolecular annihilations, thick emission layers appear favorable. When field quenching is included in addition to annihilation processes, laser operation is not achieved.

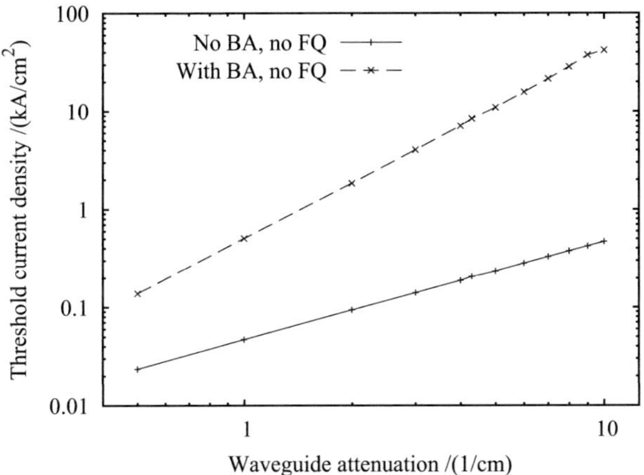

Figure 6.4: Threshold current density as a function of the waveguide attenuation, when annihilation processes are included and in their absence. Without losses, the threshold current density is a linear function of α_{WG}. In the presence of bimolecular annihilations j_{thr} follows the power law α_{WG}^2. These calculations have been carried out in the absence of field quenching.

In figure 6.4, the threshold current density j_{thr} is shown as a function of the waveguide attenuation coefficient α_{WG}. In the absence of bimolecular annihilations (BA) and field quenching (FQ), j_{thr} is a linear function of α_{WG}. When α_{WG} is decreased from 4.3 cm^{-1} to 1 cm^{-1}, j_{thr} is reduced from 206 A/cm^2 to 47 A/cm^2. In the presence of BA, j_{thr} follows the power law α_{WG}^2 for $\alpha_{\mathrm{WG}} > 1$ cm^{-1}, which is a consequence of the bimolecular character of annihilation processes. In this case, j_{thr} is reduced by a factor of 17 from 8.5 kA/cm^2 to 0.5 kA/cm^2, when α_{WG} is decreased from 4.3 cm^{-1} to 1 cm^{-1}. This result demonstrates the importance of the waveguide quality for the threshold current density.

6.2.3 Field quenching

Figure 6.5 shows the impact of bimolecular annihilations (BA) and field quenching (FQ) on the singlet exciton density $n_{\mathrm{S_1}}$ as a function of current density. If neither BA nor FQ are included, $n_{\mathrm{S_1}}$ increases linearly with current density. In this case, laser operation is achieved at a current density of 206 A/cm^2. In the presence of BA (but without FQ), $n_{\mathrm{S_1}}$ is still increasing. However, the slope is decreasing for current densities above 1 A/cm^2, pointing out a gradual reduction of the quantum efficiency. At the threshold current density of 8.5 kA/cm^2, BA reduces the singlet exciton density by about two orders of magnitude.

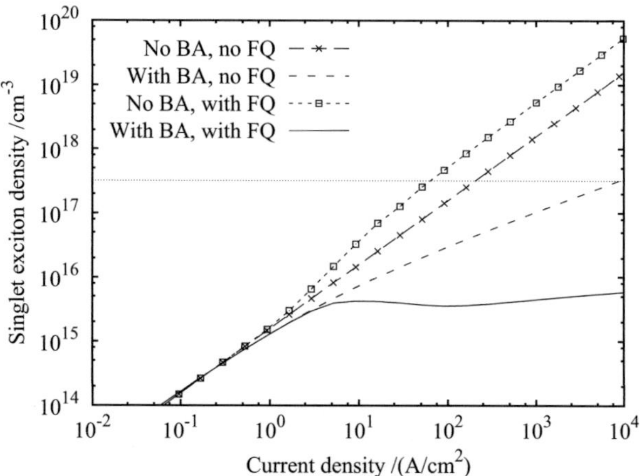

Figure 6.5: Singlet exciton density for increasing current density with and without consideration of bimolecular annihilations (BA) and field quenching (FQ). The horizontal curve indicates the singlet exciton density at the laser threshold for the considered device.

In the presence of FQ (but without BA), n_{S_1} is increased. The electric field greatly dissociates triplet excitons. These dissociated charge carriers can again undergo Langevin recombination. Depending on the splitting of the singlet and triplet exciton binding energies, the dissociation of triplet excitons dominates at high current densities.

As shown in figure 6.5, n_{S_1} is decreased by BA and FQ for current densities above 10 A/cm^2. This result is remarkable, since the singlet exciton density is only affected to a much smaller extent, when BA or FQ are separately present. This huge quenching of n_{S_1} can be understood by analyzing the polaron density as a function of current density (figure 6.6). In the presence of FQ, both singlet and triplet excitons are strongly dissociated, hence the polaron density is strongly increased. Above 1 kA/cm^2 the polaron density is increased by a factor of 10 due to FQ. Hence, the influence of singlet-polaron annihilation on n_{S_1} is considerably larger. For the employed parameters, n_{S_1} does not exceed the value of 6×10^{15} cm^{-3}, when BA and FQ are present, simultaneously. Hence, laser operation is not achieved.

6.3 Impact of device geometry

In this section we study the singlet exciton density as a function of current density for different layer thicknesses. We investigate, whether the impact of loss processes

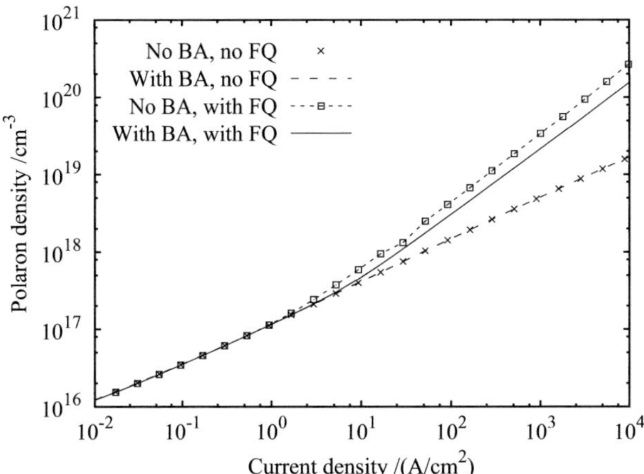

Figure 6.6: Polaron density as a function of current density when bimolecular annihilations (BA) and field quenching (FQ) are considered and their absence. In the presence of FQ, the polaron density is greatly increased. Hence, the impact of singlet-polaron annihilation on the singlet exciton density is more striking.

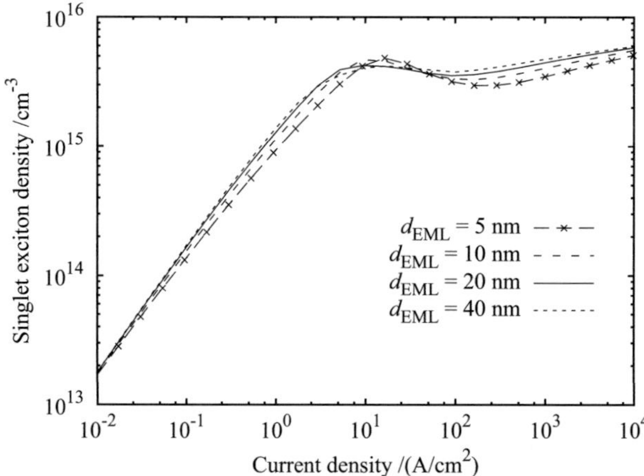

Figure 6.7: Singlet exciton density as a function of current density for various emission layer thicknesses. In the presence of annihilations and field quenching, the singlet exciton density is only increased marginally, when thicker emission layers are employed.

can be reduced by changing the thicknesses of the emission and the transport layers.

6.3.1 Emission layer thickness

The singlet exciton density n_{S_1} is depicted as a function of current density for different emission layer thicknesses in figure 6.7. Here, only the thickness of the emission layer d_{EML} is altered while transport layers remain unchanged ($d_{\mathrm{TL}} = 150$ nm). For the investigated thicknesses, the impact of d_{EML} on n_{S_1} is marginal. n_{S_1} increases by less than a factor of 2, when the emission layer is increased from 20 nm to 40 nm. For thinner emission layers, n_{S_1} is slightly decreasing, since the particle density within the emission layer is higher. Therefore, the impact of bimolecular annihilation is enhanced. For the considered layer thicknesses, the impact of losses cannot be reduced significantly by adjusting the emission layer thickness.

6.3.2 Transport layer thickness

In order to reduce the operation voltage, thinner organic layers appear favorable. For the considered TE$_2$-mode design concept, thinner devices are only favorable, if the waveguide attenuation is not increased at the same time. For materials exhibiting higher refractive indices or optical gain at shorter wavelengths, this goal could be achieved. Figure 6.8 presents the singlet exciton density n_{S_1} as a function of the

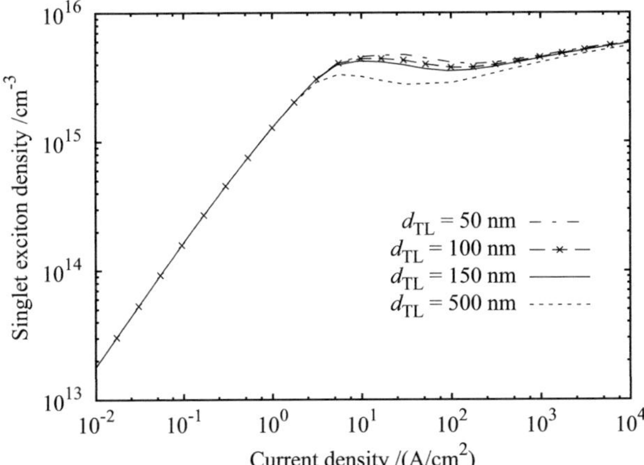

Figure 6.8: Singlet exciton density as a function of current density for various transport layer thicknesses. Here, the influence of annihilations and field quenching is take into account. The increase of the singlet exciton density is only negligible, when thinner transport layers are employed.

current density for various transport layer thicknesses. Here, the emission layer thickness remains unchanged ($d_{\mathrm{EML}} = 20$ nm). For the investigated transport layer thicknesses, the singlet exciton density is almost unaffected. $n_{\mathrm{S_1}}$ is only increased by less than a factor of 1.5 when d_{TL} is varied from 50 nm and 500 nm, since the internal electric field is not decreased for thinner transport layers. Hence, $n_{\mathrm{S_1}}$ is not increased significantly for thinner transport layers.

6.4 Impact of material properties

In this section, the impact of various material properties on the singlet exciton density is studied as a function of current density. Material parameters are determined, whose optimization may allow higher singlet exciton densities under electrical excitation. In section 6.4.1, the influence of charge carrier mobilities is analyzed. In section 6.4.2, the effect of the exciton binding energies is examined. The influence of bimolecular annihilations is studied in section 6.4.3.

6.4.1 Charge carrier mobilities

In figure 6.9, the singlet exciton density $n_{\mathrm{S_1}}$ is shown as a function of the current density for different charge carrier mobilities. For all studied values, the mobilities

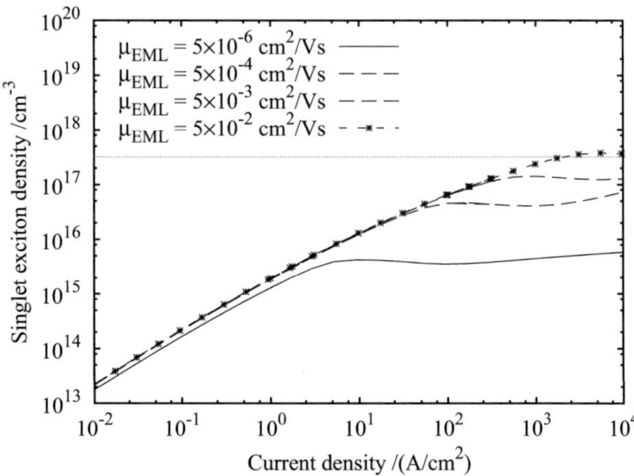

Figure 6.9: Singlet exciton density as a function of current density for various charge carrier mobilities in the emission layer. The mobilities of the transport layers are 10 times higher than the mobilities in the emission layer.

in the transport layers are 10 times higher than the mobilities in the emission layer. When the mobilities are improved, n_{S_1} can be greatly increased for high current densities. For mobilities of 5×10^{-3} cm^2/Vs, n_{S_1} reaches a value of 10^{17} cm^{-3} at a current density of 300 A/cm^2. When the current density is further increased to 10 kA/cm^2, the singlet exciton density drops by a factor of 1.5, since exciton losses by field quenching and singlet-polaron annihilation increase much faster than exciton generation by Langevin recombination. This effect would cause a bistable behavior of the electroluminescence. Optical gain would be achieved at a singlet exciton density of about 3×10^{17} cm^{-3}, which is achieved for charge carrier mobilities of 5×10^{-2} cm^2/Vs at a current density of 2 kA/cm^2.

For most organic materials used in OLED geometries, typical hole mobilities are in the range from 10^{-6} cm^2/Vs to 10^{-4} cm^2/Vs [194, 196, 220]. Typical electron mobilities are in the range from 10^{-8} cm^2/Vs to 10^{-5} cm^2/Vs [198, 221]. Recently, hole mobilities of up to 10^{-2} cm^2/Vs [157, 196] and electron mobilities of up to 10^{-3} cm^2/Vs [197] have been reported. Hence, laser operation does not seem to be feasible for the available materials.

Figure 6.10 shows the variation of the charge carrier mobility of the transport layer, while the mobility within the emission layer is kept constant. In this case, the singlet exciton density can still be increased. For mobilities in the transport layers of about 0.5 cm^2/Vs, n_{S_1} reaches a value of 10^{17} cm^{-3} at a current density of 4 kA/cm^2. The great increase of n_{S_1} mainly results from better conductivity of the

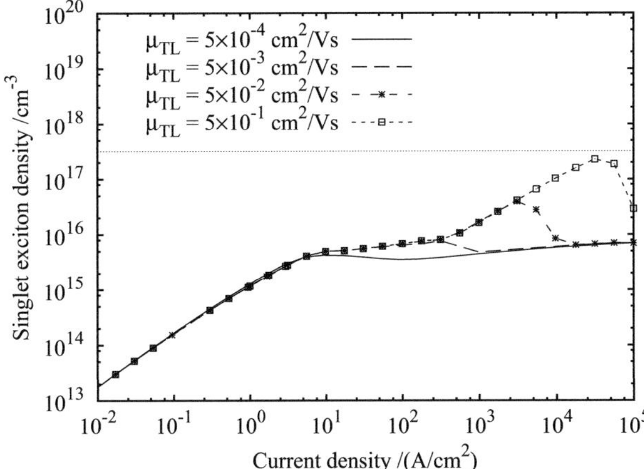

Figure 6.10: Singlet exciton density as a function of current density for various charge carrier mobilities in the transport layer.

transport layers and a lower operation voltage. This result indicates that the use of doped or crystalline materials in the transport layers might be advantageous, which would increase the conductivity within the transport layers significantly. However, spectral separation of induced absorption bands of the transport layers from the laser wavelength is essential.

6.4.2 Exciton binding energies

Within the Onsager model, the impact of field quenching is determined by the binding energies E_{S_1} and E_{T_1} of the singlet and triplet excitons, respectively. Here, the impact of E_{S_1} on the singlet exciton density n_{S_1} is analyzed. Figure 6.11 shows n_{S_1} as a function of current density for various singlet exciton binding energies. The energy splitting $E_{T_1} - E_{S_1}$ is 0.2 eV for all curves. When E_{S_1} is increased from its standard value of 0.4 eV to 1.4 eV, n_{S_1} reaches a value of 10^{17} cm^{-3}. Hence, materials exhibiting high singlet exciton binding energies are favorable for laser diodes, since the inset of field quenching occurs at higher current densities. When the current density is increased further, the accompanying increased field quenching of singlet excitons becomes the dominating loss channel and the effect of the higher binding energy is not decisive any more.

The influence of the energy splitting $E_{T_1} - E_{S_1}$ on n_{S_1} is demonstrated in figure 6.12. Here, E_{S_1} is kept constant at 0.4 eV. When $E_{T_1} - E_{S_1}$ is increased from 0.2 eV to 0.3 eV, n_{S_1} strongly decreases for current densities above 10 A/cm^2. In this case, the dissociation rate for singlet excitons is much higher than the dissociation rate for

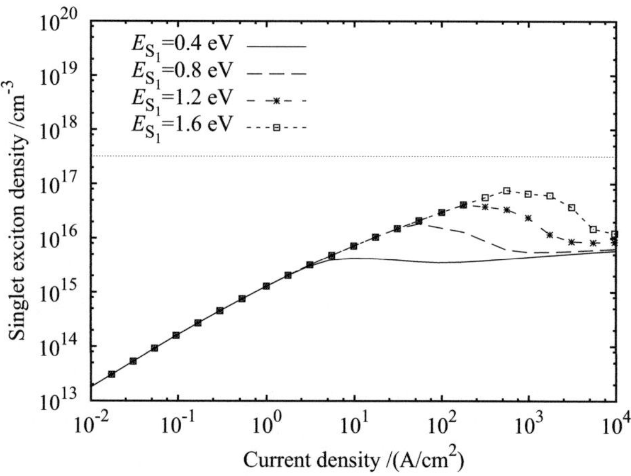

Figure 6.11: Singlet exciton density as a function of current density for various singlet exciton binding energies.

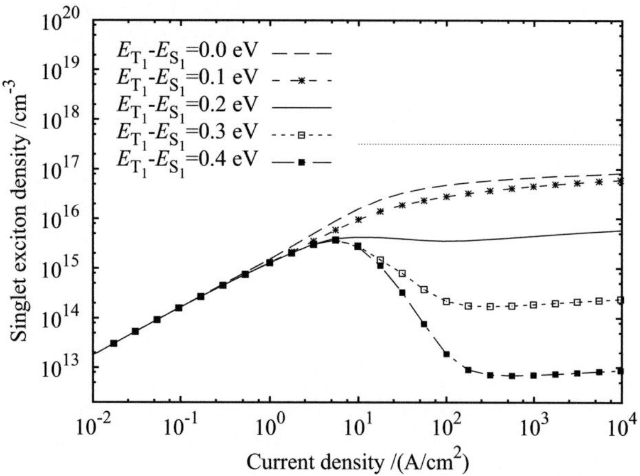

Figure 6.12: Singlet exciton density as a function of current density for various values of the energy splitting $E_{T_1} - E_{S_1}$.

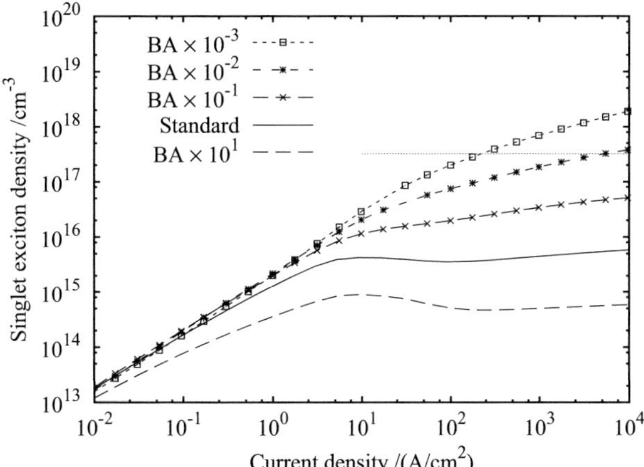

Figure 6.13: The impact of bimolecular annihilation processes on the singlet exciton density as a function of current density. The rate coefficients of each annihilation process is scaled by a fixed factor. If the annihilation rate coefficients was reduced by an order of magnitude, n_{S_1} would also increase by almost an order of magnitude at a current density of $10\ \mathrm{kA/cm^2}$.

triplet excitons. Hence, the triplet states accumulate while n_{S_1} is additionally reduced by singlet-triplet annihilation. When $E_{T_1} - E_{S_1}$ is reduced to 0 eV, n_{S_1} increases by almost two orders of magnitude. In this case, triplet excitons are dissociated more efficiently to polarons, which may again undergo Langevin recombination. Hence, materials are favorable, which offer a small splitting of the binding energies.

6.4.3 Annihilation processes

As shown in figure 6.6, the density of polarons is strongly influenced by field quenching. In this section, the role of the rate coefficients for bimolecular annihilation processes in the presence of field quenching is studied. In figure 6.13, n_{S_1} is calculated as a function of current density for different annihilation rate coefficients. In order to investigate the overall impact of annihilation processes, all rate coefficients are changed simultaneously by the same factor. If the rate coefficients are increased by a factor of 10, the singlet exciton density is reduced by an order of magnitude from $5 \times 10^{15}\ \mathrm{cm^{-3}}$ to $5 \times 10^{14}\ \mathrm{cm^{-3}}$ at a current density of $10\ \mathrm{kA/cm^2}$. If the impact of annihilation processes is weaker, the singlet exciton density increases significantly. If the rate coefficients are reduced by two orders of magnitude, the singlet exciton density exceeds $10^{17}\ \mathrm{cm^{-3}}$ and optical amplification would be possible. Hence, materials

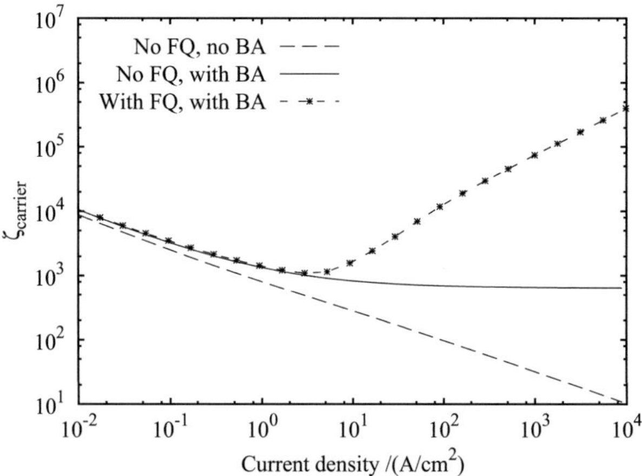

Figure 6.14: Polaron absorption parameter ζ_{carrier} as a function of current density is shown when bimolecular annihilations (BA) and field quenching (FQ) is considered and in their absence.

exhibiting low rate coefficients for bimolecular annihilation, are preferable and laser operation would be possible even if field quenching is taken into account. This study also illustrates, that the knowledge of the rate coefficients for all loss mechanisms is a prerequisite for the careful selection of suitable materials for an organic laser diode.

6.5 Implications for induced absorption processes

Figure 6.14 shows the polaron absorption parameter ζ_{carrier} as a function of current density. Without bimolecular annihilations (dashed curve), ζ_{carrier} follows the power law $j^{-0.5}$ for all shown current densities, which is a consequence of space charge limited current and Langevin recombination. Hence, for any set of cross sections for stimulated emission and polaron absorption, ζ_{carrier} can be reduced without limitations by increasing the pump power. Therefore, laser operation is possible in all cases.

For BA being included (solid curve), singlet excitons are also quenched by bimolecular annihilation processes like singlet-singlet and singlet-polaron annihilation. These processes become increasingly important at high current densities causing ζ_{carrier} to saturate at 650 for current densities above 10^3 A/cm^2.

In the presence of BA and FQ (dashed curve with crosses), ζ_{carrier} is first decreasing for current densities below 5 A/cm^2. At $j = 5$ A/cm^2, ζ_{carrier} reaches its minimum value of 1150. For higher current densities, ζ_{carrier} increases again. At the threshold

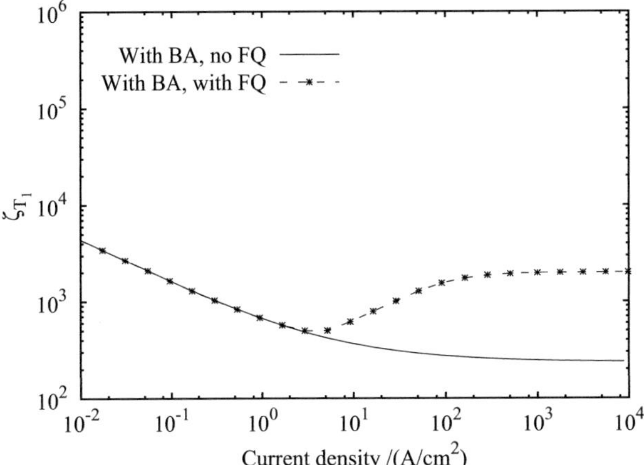

Figure 6.15: Triplet absorption parameter ζ_{T_1} as a function of current density is shown when bimolecular annihilations (BA) and field quenching (FQ) is considered and in their absence.

current density in the absence of induced absorption $j_{thr} = 8.5$ kA/cm^2, $\zeta_{carrier}$ is 3.4×10^5. Hence, for the employed parameters, laser operation will be prohibited, if $\sigma_{carrier} > 3 \times 10^{-6} \times \sigma_{SE}$. Hence, laser operation can only be achieved in the presence of BA and FQ, if the polaron absorption band can be separated spectrally from the laser wavelength and the stimulated emission cross section is at least 6 orders of magnitude higher than the polaron absorption cross section.

In figure 6.15, ζ_{T_1} increases with increasing current density. In the presence of bimolecular annihilation, ζ_{T_1} saturates at a value of 237 for current densities above 5 A/cm^2. In the presence of field quenching, ζ_{T_1} saturates at about 2010. Hence, electrically pumped laser operation will be prohibited, if $\sigma_{T_1 T_N} > 5 \times 10^{-4} \times \sigma_{SE}$.

6.6 Guest-Host-Systems

Recently, the impact of FQ has been studied in guest-host systems such as doped small molecule materials [133]. In this paper, the effect of field quenching is examined for dye-doped Alq$_3$ films and their field-dependent photoluminescence efficiency is compared to the efficiency in pristine Alq$_3$ layers. In figure 6.16, the measured photoluminescence efficiency is plotted for increasing electric field for pristine Alq$_3$layers (open circle) and 4-dicyanomethylene-2-t-butyl-6-1,1,7,7-tetra methyljulolidyl-9-enyl-4H-pyran (DCJTB) doped Alq$_3$ layers (solid circles) [133]. At an electric field of 1.5 MV/cm, the photoluminescence efficiency drops below 70% in undoped layers, while the efficiency is less affected for dye-doped Alq_3 films. For DCJTB doped Alq_3 layers, the quenching efficiency is reduced to less than 10%. The good agreement between experimental data and the hopping separation model, which has been discussed in section 3.4.2.4, is achieved for the field-dependent PL efficiency of pristine Alq_3 (solid curve) and DCJTB doped Alq_3 (dashed curve).

The authors suggest that the reduction the field quenching rate is attributed to the energy band gap of the guest molecule, which is narrower than the one of the host material. Therefore, it is less affected by field-induced exciton dissociation. For higher doping concentrations and for smaller band gaps, the impact of FQ is further decreased.

In figure 6.17, the singlet exciton density is shown as a function of the current density, when the best fit for Alq_3 and DCJTP doped Alq_3 are used as field quenching rates. For current densities below 10 A/cm^2, the exciton densities of Alq_3 and DCJTB doped Alq_3 are virtually identical. However, field quenching greatly reduces the exciton density in the Alq_3 layer. The simultaneous impact of bimolecular annihilation processes and field quenching cause a strong reduction of the exciton density in Alq_3 for current densities above 10 A/cm^2. In contrast, the impact of field quenching is greatly reduced in for DCJTB doped Alq_3. At a current density of 10 kA/cm^2, a singlet exciton density of 10^{17} cm^{-3} is derived, which is of the same order of magnitude as the exciton densities in optically pumped low-threshold organic lasers. Thus, in an actual laser diode, the impact of field quenching in the active material should be equal to or smaller than the impact of field quenching in Alq_3:DCJTB.

6.7 Chapter summary

Besides bimolecular annihilation processes and induced absorptions by polarons and triplet excitons, field-induced exciton dissociation (field quenching) is found to be another obstacle towards the realization of an organic injection laser. In this section, the impact of field quenching has been studied for a double heterostructure laser diode by numerical simulation. Laser operation can still be achieved, if bimolecular annihilations or field quenching are present separately. When both loss mechanisms are present simultaneously, laser operation is prohibited for all current densities for the employed parameters. For current densities above 1 A/cm^2, the polaron density is strongly increased in the presence of field quenching. When bimolecular annihilations are included simultaneously, the impact of singlet-polaron annihilation is strongly enhanced and the singlet exciton density does not exceed 6×10^{15} cm^{-3} for current densities below 10 kA/cm^2.

The singlet exciton density remains almost unaffected, when the thicknesses of the emission and the transport layers are altered. A low attenuation coefficient of the waveguide appears crucial for lasing.

An enhancement of the charge carrier mobilities well above the current values reported for organic LED geometries is a prerequisite for electrically induced laser operation. Mobilities of about 0.5 cm^2/Vs in the emission layer and 5 cm^2/Vs in the

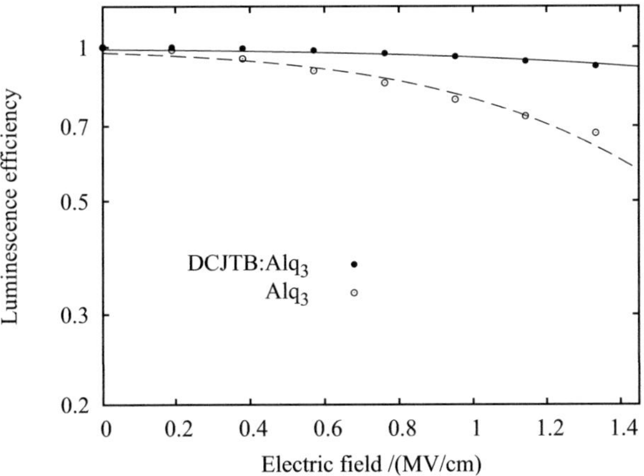

Figure 6.16: Photoluminescence efficiency for increasing electric field for pristine Alq$_3$layers (circle) and DCJTB doped Alq$_3$ (cross) [133]. The experimental values are fitted using an Onsager model.

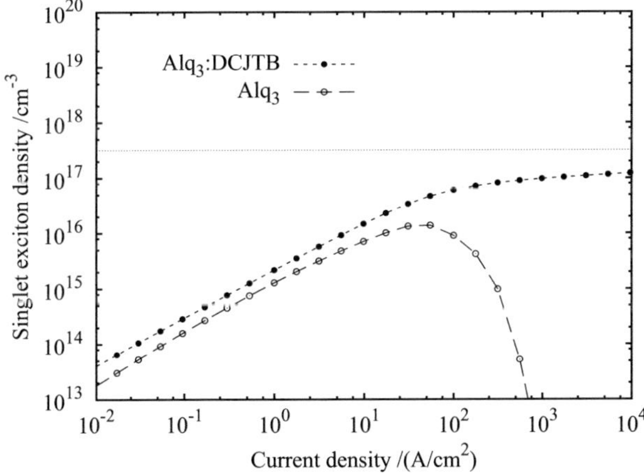

Figure 6.17: Photoluminescence efficiency for increasing electric field for pristine Alq$_3$layers (circle) and DCJTB doped Alq$_3$ (cross). The experimental values are fitted using an Onsager model.

transport layers are necessary in order to increase the singlet exciton density beyond the laser threshold at about 10^{17} cm^{-3}. Additionally, guest-host systems seem so be favorable. Here, the energy band gap of the guest molecule is narrower than the one of the host material and excitons are less affected by field-induced exciton dissociation [133].

The interplay of bimolecular annihilation and field quenching has been analyzed. For bimolecular annihilation, small rate coefficients are favorable. Regarding field quenching, materials should be preferred, exhibiting firstly high singlet exciton binding energies and secondly small splitting between the binding energy for singlet and triplet excitons.

The impact of polaron and triplet absorption is further increased in the presence of field quenching. These induced absorption processes may prohibit laser operation for all current densities. For optical gain, the polaron absorption cross section has to be at least 3×10^{-6} times smaller than the cross section for simulated emission. Additionally, the cross section for simulated emission has to exceed the triplet-triplet absorption cross section by almost 4 orders of magnitude.

In this chapter, organic laser diode structures under high excitation have been studied by numerical simulation. The impact of bimolecular annihilations, field-induced exciton dissociation and polaron as well as triplet absorption on the threshold current density has been studied. Two design concepts have been evaluated, which are promising concepts for an organic laser diode. The best performance is achieved for a TE_2-mode device structure, which combines low waveguide attenuation as well as thin organic layers. When all loss processes are included, a singlet exciton density of 10^{17} cm^{-3} for DCJTB doped Alq$_3$ at a current density of 10 kA/cm^2 is derived, which is of the order of magnitude of the lowest laser thresholds for optical pumping.

7 Design rules for organic laser diodes

In the previous chapters, we have analyzed the impact of losses on the threshold current density of organic laser diodes. In this section, the elaborated design rules for an organic laser diode are briefly summarized.

TE$_2$-mode waveguide The absorption of the waveguide has to be minimized below 5 /cm, while at the same time, the total thickness of the organic layers has to be as low as possible in order to reduce the operating voltage. This trade-off is fulfilled to the best extent for the TE$_2$-structure. Waveguide attenuation coefficients of 4.3 /cm have been demonstrated experimentally [167] for a the total device thickness of 320 nm. Therefore, we suggest that an organic laser diode should exhibit the TE$_2$-geometry.

Balanced and high charge carrier mobilities In order to reduce the impact of loss processes under high excitation, the charge carrier mobility has to be increased. By increasing the charge carrier mobility, the impact of all loss processes is reduced. Due to the lower particle density, the impact of bimolecular annihilations and induced absorption losses is reduced. In addition, the electric field is decreased for materials with higher charge carrier mobilities. Hence the impact of field-induced exciton dissociation is lower. Our calculations suggest, that for an optimized design, charge carrier mobilities of at least 5×10^{-2} cm^2/Vs are required in order to achieve laser operation at a current density of about 10 kA/cm^2.

However, high charge carrier mobilities are not sufficient in order to achieve laser operation. Additionally, the injection of electrons and holes has to be balanced in order to achieve homogeneous exciton generation in the emission layer. For unbalanced charge carrier mobilities, excitons are generated in a narrow area, since the charge carrier species, which exhibits the lower mobility, will only be injected a few nanometers into the emission layer.

Emission layer thickness For thicker emission layers, the filling factor of the laser mode in the emission layer is increased. If charge carriers are still balanced within the emission layer (which means, that excitons are generated all over the emission layer), a smaller exciton density is required in order to achieve laser operation. Thus, the impact of annihilation processes is reduced, since the densities of polarons and excitons are smaller. From the technological point of view, an emission layer of a thickness of 20 nm is feasible, which still exhibits the abovementioned requirements.

For thicker emission layers, the threshold current density would be decreased due to a reduced impact of bimolecular annihilation processes.

Materials exhibiting low bimolecular annihilations   The rate coefficients of bimolecular annihilation processes can vary by up to two orders of magnitude for different materials. Since the impact of annihilation processes on the threshold current density is tremendous, materials exhibiting low annihilation rate coefficients should be used in the emission layer. Singlet polaron annihilation has been identified as the most striking process.

Guest-host systems   In the emission layer, guest-host systems should be used. On the one hand, guest-host systems exhibit low self-absorption. Thus, the waveguide attenuation for the laser mode can be reduced. On the other hand, the impact of field-induced exciton dissociation can be reduced significantly for materials, where a low-bandgap guest is doped into a high bandgap host. Our calculations show that even in the presence of field quenching, a sufficiently high singlet exciton density can be achieved in guest-host systems.

Pulsed operation   In order to achieve current densities of several $100\,\mathrm{A/cm^2}$, pulsed excitation is essential in order to avoid thermal destruction of the device. In addition, the impact of triplet absorption can be reduced by pulsed operation, since the triplet excitons reach their steady-state density on a longer timescale than singlet excitons. Reverse current pulses, which extract polarons from the device, might be used in order to reduce the impact of polaron absorption. However, this approach is technologically challenging, since rise times of about 2 ns have to be achieved for the reverse pulse in order to avoid exciton losses due to spontaneous decay.

8 Conclusions and outlook

Organic semiconductor lasers exhibit unique features such as the tunability over the whole visible range, low-cost production by using printing techniques and the fabrication on flexible substrates. Due to their special properties, organic semiconductor lasers are suitable for a number of applications in bioanalytics, digital printing and security applications. Until today, all organic semiconductor lasers are exclusively optically pumped by an additional laser source, such as a gas laser, inorganic laser diode or an inorganic light emitting diode. Since inorganic laser sources are much more expensive than the actual organic laser, the unit cost is mainly determined by the cost of the additional pumping source. Thus, several applications are not feasible using optically pumped organic lasers, where the unit cost has to be of the order of a few cents. Electrically pumped organic lasers would exhibit the potential to achieve this target, since a second laser source is not required.

When compared to optically pumped systems, electrically excited organic semiconductor lasers are much more challenging. Firstly, highly conductive contact electrodes have to be integrated into a low-loss laser waveguide. Secondly, current densities of at least several 100 A/cm^2 have to be injected in order achieve sufficient excited states for optical gain. Since the charge carrier mobilities in organic semiconductors are much lower than in their inorganic counterparts, charge carrier densities above 10^{19} $/cm^3$ and electric fields of several MV/cm are necessary for current densities of 100 A/cm^2. Thus, induced loss processes become important, which strongly increase the threshold current density.

In this work, the feasibility of an organic laser diode has been evaluated by numerical simulations. A one-dimensional drift-diffusion model is employed, which describes the electrical properties of the organic multilayer device. The temporal and spatial distributions of the particle densities and the electric field are calculated. The optical properties are described by using a transfer-matrix method. Loss processes are included, which occur at high excitation and whose impact is in particular striking for organic laser diodes. The impact of bimolecular annihilations, polaron absorption, triplet absorption as well as field-induced exciton dissociation are considered. For the considered devices, the threshold current densities are calculated. The impact of the device geometry on the threshold current density is studied in order to derive design rules for device architectures. The influence of many material parameters on the threshold current density is evaluated in order to derive selection criterions for proper laser materials.

Organic double-heterojunction devices with different waveguide designs and electrode concepts have been evaluated. The best results have been obtained for a TE_2-mode design concept, which combines a low-attenuation waveguide and thin organic layers with a total thickness of 320 nm. If no loss mechanisms are considered, a threshold current density of 206 A/cm^2 is derived for the considered device. In the presence of bimolecular annihilation processes, the threshold current density increases to 8.5 kA/cm^2. When the impact of annihilations and the impact of field-induced exciton dissociation is considered simultaneously, laser operation is not achieved. Our simulations suggest that field-induced exciton dissociation increases the polaron density in the emission layer whereby the impact of singlet-polaron annihilation is strongly increased. However, the impact of field-induced exciton dissociation is decreased in guest-host systems, where a low-bandgap guest is doped into a high-bandgap host.

Device design rules for laser diode structures are elaborated and material parameters are identified, whose optimization may decrease the impact of field quenching, significantly. The singlet exciton binding energy and the energy splitting between the binding energies of singlet and triplet excitons are found to be decisive material parameters. Increasing the conductivity of the transport layers substantially decreases the impact of field quenching, even if the charge carrier mobility in the emission layer remains unchanged. Furthermore, the simultaneous action of annihilation processes and field quenching is illustrated. Regarding induced absorption processes, maximum relative cross sections of 3×10^{-6} for polarons and 5×10^{-4} for triplet excitons have been calculated, which would still allow laser operation. For DCJTB doped Alq$_3$, a singlet exciton density of about 10^{17} cm^{-3} would be achieved at a current density of 10 kA/cm^2, which is of the same order of magnitude as singlet exciton densities in optically pumped low-threshold organic lasers.

In conclusion, electrically pumped lasing in organic semiconductors is only possible for materials, which fulfill a very small parametric window. Besides the design concepts, which have been evaluated in this thesis, further device designs are currently evaluated. Organic double-heterojunction field-effect transistors enable an increased charge carrier mobility, which would decrease the impact of charge carrier related losses. Alternatively, capacitively pumped organic light emitting diodes, which are operated under alternating current, are currently investigated. These devices allow a further separation of the electrodes from the waveguide and might enable even thinner transport layers and lower waveguide attenuation coefficients.

9 Acknowledgement

This work would not have been possible without the support of many individuals.

- First of all, I would like to thank my advisor Uli Lemmer for the opportunity to work in his group on a very interesting and exciting topic during the last three years as well as for the excellent scientific environment. I really enjoyed many fruitful discussions and I greatly benefited from his enormous expertise in organic materials.

- I thank Wolfgang Kowalsky for his interest in my work and for the commitment to officiate the official report, i.e. für die Übernahme des Korreferats.

- All collaborators within the BMBF-project "Organische Laserdiode": Patrick Görrn, Torsten Rabe, Thomas Riedl, Wolfgang Kowalsky, Uli Scherf, Bodo Wallikewitz, Dirk Hertel, Klaus Meerholz, Thomas Weimann, Hartmut Fröb and Karl Leo.

- I would like to thank my colleagues Jan Brückner, Marc Stroisch, Christian Karnutsch, Sebastian Valouch and Thomas Woggon for their experimental work on organic semiconductor lasers as well as for many fruitful discussions on organic optoelectronic devices and on concepts for organic laser diodes.

- I thank Christof Pflumm for his patient support with the SONDE source code.

- Felix Glöckler and Martina Gerken for their competent support with the optical modelling of organic devices.

- I am grateful to Yousef Nazirizadeh for his support with the transient photo-luminescence measurement setup.

- I greatly appreciated to work with Nico Christ and Stephan Uebe during their Diploma work at the LTI.

- I am grateful to Christian Kayser, Alexander Colsmann, Thorsten Feldmann and Boris Riedel for the preparation of organic devices and their technological support.

- All colleagues and former colleagues at the Light Technology Institute for the excellent collaboration and working atmosphere.

9 Acknowledgement

List of Figures

2.1 Electron configuration of carbon: ground state (top left), sp^3- (top right), sp^2- (down left) and sp-hybridization (down right). For the ground state, the 1s- and the 2s- orbitals are occupied by two electrons and two 2p-orbitals are occupied by one electron. Hybrid orbitals are formed by superposition of s- and p-orbitals. 7

2.2 For ethane, carbon has four single bonds (sp^3-hybridization). For ethene, the two carbon atoms are bound by a double bond (sp^2-hybridization). Here, the 2p-orbitals form a π-bond. For ethyne, carbon is even able to form a triple bond (sp-hybridization) which consists of a σ-bond and two π-bonds. 7

2.3 Chemical structure of polyethyne which is a basic example for a conjugated polymer. The electrons of the π-bond are weakly bound and delocalized along the polymer backbone. The π-electrons can be excited by visible light and electric charge transport is possible. 8

2.4 Chemical structure of conjugated polymer materials: poly(9,9-dioctylfluorene) (PFO) (top left), polyphenylene vinylene (PPV) (top right), poly(9,9-dioctylfluorene-co-benzothiadiazole (F8BT) (down left) and ladder-type polyparaphenylene (MeLPPP) (down right) 9

2.5 Chemical structures for the small molecule materials aluminum tris(8-hydroxyquinoline) (Alq$_3$), 4,7-Diphenyl1,10-phenanthroline (Bphen), N,N'-di(naphthalene-1-yl)N,N'-diphenylbenzidine (NPB) and N,N'-DiphenylN,N'-bis(m-tolyl)benzidin (TPD). These materials are used in organic light-emitting diodes (OLEDs) and organic lasers. 10

2.6 Conjugated dendrimer (schematically) consisting of conjugated units (C), linkers (L) and surface groups (S). The optical properties of the molecular core can be tuned independently from the electrical properties of the surrounding branches. 11

2.7 Energy diagram of the excitonic ground state S$_0$ and the first excited singlet state S$_1$. Under optical excitation, organic semiconductors are absorbing for photon energies above the bandgap energy. The absorbing molecule undergoes a transition from the ground state S$_0$ to a vibronic substate of the S$_1$ band. Following this, the molecule thermalizes with a lifetime of 1 ps to the lowest vibronic state. From the S$_1$ state, the molecule decays radiatively with a typical lifetime of 1 ns. Due to the relaxation mechanism, the photoluminescence (PL) is red-shifted relatively to the absorption (Stokes-shift). 13

2.8   Jablonski-diagram of an organic molecule showing the basic transitions between the ground state S_0 and the excited singlet and triplet states S_1, S_2, T_1 and T_2. For details see text. 14

2.9   Schematic energy diagram of polaronic levels. New polaronic states are formed within the bandgap. In organic laser diodes, additional induced absorptions are caused by polarons. 16

2.10   Charge transport in organic semiconductors is based on hopping between localized states. The mobility is determined by the distribution of distances between sites and their energetic distribution. 17

2.11   Within the Poole-Frenkel model, charge transport is modeled by field-induced escape of charges from lower lying states below the conduction band, which are modeled as coulomb wells. 17

2.12   Carrier injection from a metal contact into an organic semiconductor. 20

2.13   Carrier injection from a metal contact into an organic semiconductor by Fowler-Nordheim tunneling. 21

2.14   Schematic illustration of Förster energy transfer. Left: An excited molecule (donor) is deactivated and its energy is transferred to an accepting molecule. Right: The absorbing energies of the acceptor have to match the emitting energies donor. 24

2.15   Schematic illustration of Förster energy transfer. An excited molecule (donor) is deactivated and its energy is transferred to an accepting molecule. 25

2.16   Schematic illustration of Dexter transfer. The energy is transferred by exchange of a charge pair. Thus, only the complete spin of the donor and the acceptor has to be conserved. Therefore, the transfer of singlet and triplet excitons is allowed. 26

2.17   Overview of bimolecular annihilation processes in organic semiconductors. Most annihilation processes quench singlet excitons, being the species responsible for optical gain. 27

2.18   Schematic illustration of polaron absorption in organic semiconductor materials. For details see text. 32

2.19   Schematic illustration of triplet absorption in organic semiconductor materials. For details see text. 33

2.20   Schematic illustration of field-induced dissociation of singlet and triplet excitons. 34

2.21   Schematic illustration of a laser, consisting of an amplifying material and two mirrors, which form a resonator. 38

2.22   Left: Schematic illustration of a four-level laser system. Radiative transitions are indicated by thick arrows. Right: Energy scheme of an organic laser material. For details see text. 38

2.23   Schematic illustration of an organic distributed feedback laser which consists of a structured glass substrate and an organic semiconductor as the active material. Optical feedback is generated by the periodic modulation of the refractive index. 39

2.24 Schematic illustration of an organic laser diode structure based on an OLED design: Organic double-heterostructure with thick transport layers for propagation of a TE_0 laser mode. 41

2.25 An organic laser diode structure based on an OLED design, which is designed for propagation of a TE_2-mode (schematic illustration, adapted from [167]). This design combines thin organic transport layers and a low attenuation coefficients of the waveguide. 42

2.26 Lateral view of an ambipolar organic field-effect transistor (schematic). 44

2.27 Lateral view of an organic light emitting diode with field-enhanced electron transport (adapted from [162]). 45

2.28 Geminate polaron pair accumulation device structure consisting of a coumarin 6 (C6) doped 4,4-bisN-carbazolylbiphenyl (CBP) emission layer, which is sandwiched between two Teflon AF^{TM} layers (adapted from [190, 191]). 46

2.29 Energy band scheme for the guest-host system of C6 doped CBP (adapted from [190, 191]). The hole is trapped on the C6 while the electron is mobile within the attractive potential of the hole. 46

3.1 Device geometry for a double-heterostructure organic laser diode, consisting of an electron transport layer (ETL), emission layer (EML) and a hole transport layer (HTL). Each layer is specified by a one dimensional grid. At each grid point, the time dependent local field, particle densities, currents and reaction rates are calculated. 50

3.2 Sketch of waveguiding in a slab waveguide structure consisting of a high-index core (n_{core}) and a low-index cladding ($n_{cladding}$). Waveguiding is achieved by total reflection, if the angle of incidence Θ_m is larger than the critical angle for total reflection Θ_c. 51

3.3 Schematic lateral view of a waveguide within an organic laser diode. For low-attenuation waveguiding, the refractive index of the emission layer has to exceed the refractive index the transport layers. 52

3.4 A transfer matrix method is employed in order to calculate the modal intensity profile of the laser modes. 53

4.1 Structure of the simulated three layer laser diode with hole transport (HTL), emission (EML) and electron transport layer (ETL). 74

4.2 Threshold current density as a function of the emission layer thickness with annihilation processes and in their absence. The total device thickness is 1300 nm. 76

4.3 Total S_1 annihilation rate density as a function of position in the device for different emission layer thicknesses. 77

4.4 Contribution of the various S_1 annihilation processes to the total S_1 annihilation rate. 77

4.5   Threshold current density as a function of emission layer thickness for different singlet-singlet annihilation rate coefficients. The standard rate coefficient for SSA is indicated in the key of the figure with an asterisk. The curves for $\kappa_{\mathrm{SS}} = 10^{-10}$ cm^3/s and for $\kappa_{\mathrm{SS}} = 0$ are on top of each other.   79

4.6   Threshold current density as a function of the rate coefficient for different annihilation processes at a constant emission layer thickness of $d_{\mathrm{EML}} = 400$ nm.   80

4.7   Threshold current density as a function of emission layer thickness for different singlet-triplet annihilation rate coefficients.   80

4.8   Threshold current density as a function of emission layer thickness for different singlet-polaron annihilation rate coefficients.   81

4.9   Threshold current density as a function of emission layer thickness for different triplet-triplet annihilation rate coefficients.   82

4.10 Threshold current density as a function of emission layer thickness for different triplet-polaron annihilation rate coefficients.   83

4.11 *Total effective rate* (see equation (4.1)) describing the number of singlet excitons created (or quenched) by triplet excitons in the entire device, shown as a function of current density for different STA rate coefficients.   84

5.1 ζ_{carrier} as a function of current density with standard annihilations (solid line) and in their absence (dashed line). At current densities below 5 A/cm^2 the lines are virtually on top of each other. With annihilations, ζ_{carrier} is saturating at a value of 650 (dotted line) for current densities above 10^3 A/cm^2 while in the absence of annihilations, ζ_{carrier} follows the power $j^{-0.5}$ for all shown current densities.   88

5.2   The spatial distribution of the polaron and exciton densities and the modal intensity profile as a function of the position in the device at a current density of 10^4 A/cm^2.   89

5.3 ζ_{carrier} (solid line) and $\zeta_{\mathrm{T_1}}$ (dashed line) as a function of current density. For current densities above 10^3 A/cm^2, ζ_{carrier} and $\zeta_{\mathrm{T_1}}$ saturate at a value of 650 and 230, respectively (dotted lines).   90

5.4 ζ_{carrier} as a function of current density for various stimulated and absorption cross sections. For (2) to (5) the cross section for stimulated emission of F8BT ($\sigma_{\mathrm{SE}} = 7 \times 10^{-16}$ cm^2) has been assumed. (1) $\Sigma_{\mathrm{carrier}} = \Sigma_{\mathrm{T_1 T_N}} = \sigma_{\mathrm{SE}} = 0$, (2) $\Sigma_{\mathrm{carrier}} = \Sigma_{\mathrm{T_1 T_N}} = 0$, (3) $\Sigma_{\mathrm{carrier}} = \Sigma_{\mathrm{T_1 T_N}} = 10^{-4}$, (4) $\Sigma_{\mathrm{carrier}} = \Sigma_{\mathrm{T_1 T_N}} = 5 \times 10^{-4}$, (5) $\Sigma_{\mathrm{carrier}} = \Sigma_{\mathrm{T_1 T_N}} = 10^{-3}$. For (2) - (5) the curves show a sharp bend at the laser threshold.   91

5.5 Threshold current density j_{thr} as a function of relative polaron absorption cross section Σ_{carrier} (solid line) and as a function of the relative triplet-triplet absorption cross section $\Sigma_{\text{T}_1\text{T}_\text{N}}$ (dashed line). The saturation values $\zeta^{-1}_{\text{carrier,sat}}$ and $\zeta^{-1}_{\text{T}_1\text{T}_\text{N},\text{sat}}$ are indicated by the two vertical dotted lines. 93

5.6 ζ_{carrier} as a function of current density for a variation of the emission layer thickness d_{EML}. For current densities below 10^3 A/cm^2, devices with a thin emission layer show a smaller value of ζ_{carrier} than devices with a thicker emission layer. If the current density is larger than 10^4 A/cm^2, ζ_{carrier} saturates at about the same value of 650 for all investigated emission layer thicknesses. 95

5.7 ζ_{carrier} as a function of the current density for a variation of d_{TL}. For all current densities, devices with thin transport layers have almost the same value for ζ_{carrier} as devices with a thick transport layer. For current densities above 10 kA/cm^2, ζ_{carrier} saturates at about the same value of 650 for all investigated transport layer thicknesses. 97

5.8 ζ_{carrier} as a function of current density for various mobilities in the transport layer. For all current densities the curves are virtually on top of each other. Hence, ζ_{carrier} is not reduced for increasing transport layer mobilities μ_{TL}. 98

5.9 ζ_{carrier} as a function of current density for various mobilities in the emission layer, where the electron and hole mobilities are assumed to be equal. ζ_{carrier} is reduced by more than one order of magnitude for μ_{EML} being increased from 5×10^{-6} cm^2/Vs to 5×10^{-4} cm^2/Vs. . . . 99

5.10 ζ_{carrier} is shown as a function of the current density for various mobilities in the emission and in the transport layer, whereby $\mu_{\text{TL}} = 10\times\mu_{\text{EML}}$. By increasing the mobilities to $\mu_{\text{TL}} = 2$ cm^2/Vs and $\mu_{\text{EML}} = 0.2$ cm^2/Vs, ζ_{carrier} is reduced to 1 for current densities above 10^3 A/cm^2. 100

5.11 ζ_{carrier} as a function of the current density for various field dependence coefficients F_0 of the mobility in the emission and the transport layer. In the case of field dependent mobilities, ζ_{carrier} does not saturate at high current densities. 101

5.12 ζ_{carrier} as a function of current density for various singlet exciton lifetimes τ_{S_1}. By increasing the singlet exciton lifetime, the saturation value of ζ_{carrier} is reached already at lower current densities. 102

5.13 ζ_{T_1} as a function of current density for various triplet exciton lifetimes τ_{T_1}. By reducing the monomolecular triplet exciton lifetime τ_{T_1} to 1 ns, ζ_{T_1} can be reduced to almost 1. 103

5.14 Modal gain g_{SE}, polaron absorption parameter ζ_{carrier} and triplet absorption parameter ζ_{T_1} as a function of time. The switch-on process is specified by the characteristic times $t_{g,0.9}$ (circle), $t_{\text{carrier},2}$ (square) and $t_{\text{T}_1,0.5}$ (triangle). 106

5.15 The triplet absorption parameter for CW operation $\zeta_{\mathrm{T_1,CW}}$ and the triplet absorption parameter for pulsed operation $\zeta_{\mathrm{T_1,pulsed}} = \zeta_{\mathrm{T_1}}(t_{g,0.9})$ as a function of the charge carrier mobility in the emission layer μ_{EML}. For mobilities of $5 \times 10^{-4}\,\mathrm{cm^2/Vs}$, $\zeta_{\mathrm{T_1}}$ can be reduced to $1/60$ of its CW value. 107

5.16 The triplet absorption parameters for CW and pulsed operation $\zeta_{\mathrm{T_1,CW}}$ and $\zeta_{\mathrm{T_1,pulsed}}$ as a function of the triplet exciton lifetime $\tau_{\mathrm{T_1}}$. 108

5.17 Transient characteristics of g_{SE}, ζ_{carrier} and $\zeta_{\mathrm{T_1}}$, when a current step of $30\,\mathrm{A/cm^2}$ is applied. 109

5.18 g_{SE}, ζ_{carrier} and $\zeta_{\mathrm{T_1}}$ as a function of time, when a current pulse of $30\,\mathrm{A/cm^2}$ is applied, which is followed by a reverse current pulse of $-240\,\mathrm{A/cm^2}$. 110

6.1 The simulated double heterostructure laser diode with hole transport layer (HTL), emission layer (EML) and electron transport layer (ETL). 114

6.2 Modal intensity profile and refractive index of the employed materials as a function of position. The employed materials, refractive indices and layer thicknesses are specified in table 3.1. 115

6.3 Laser threshold current density as a function of the emission layer thickness, when annihilation processes are included and in their absence. In the presence of bimolecular annihilations, thick emission layers appear favorable. When field quenching is included in addition to annihilation processes, laser operation is not achieved. 116

6.4 Threshold current density as a function of the waveguide attenuation, when annihilation processes are included and in their absence. Without losses, the threshold current density is a linear function of α_{WG}. In the presence of bimolecular annihilations j_{thr} follows the power law α_{WG}^2. These calculations have been carried out in the absence of field quenching. 117

6.5 Singlet exciton density for increasing current density with and without consideration of bimolecular annihilations (BA) and field quenching (FQ). The horizontal curve indicates the singlet exciton density at the laser threshold for the considered device. 118

6.6 Polaron density as a function of current density when bimolecular annihilations (BA) and field quenching (FQ) are considered and their absence. In the presence of FQ, the polaron density is greatly increased. Hence, the impact of singlet-polaron annihilation on the singlet exciton density is more striking. 119

6.7 Singlet exciton density as a function of current density for various emission layer thicknesses. In the presence of annihilations and field quenching, the singlet exciton density is only increased marginally, when thicker emission layers are employed. 120

6.8 Singlet exciton density as a function of current density for various transport layer thicknesses. Here, the influence of annihilations and field quenching is take into account. The increase of the singlet exciton density is only negligible, when thinner transport layers are employed. 121

6.9 Singlet exciton density as a function of current density for various charge carrier mobilities in the emission layer. The mobilities of the transport layers are 10 times higher than the mobilities in the emission layer. 122

6.10 Singlet exciton density as a function of current density for various charge carrier mobilities in the transport layer. 123

6.11 Singlet exciton density as a function of current density for various singlet exciton binding energies. 124

6.12 Singlet exciton density as a function of current density for various values of the energy splitting $E_{T_1} - E_{S_1}$. 124

6.13 The impact of bimolecular annihilation processes on the singlet exciton density as a function of current density. The rate coefficients of each annihilation process is scaled by a fixed factor. If the annihilation rate coefficients was reduced by an order of magnitude, n_{S_1} would also increase by almost an order of magnitude at a current density of 10 kA/cm^2. 125

6.14 Polaron absorption parameter ζ_{carrier} as a function of current density is shown when bimolecular annihilations (BA) and field quenching (FQ) is considered and in their absence. 126

6.15 Triplet absorption parameter ζ_{T_1} as a function of current density is shown when bimolecular annihilations (BA) and field quenching (FQ) is considered and in their absence. 127

6.16 Photoluminescence efficiency for increasing electric field for pristine Alq$_3$layers (circle) and DCJTB doped Alq$_3$ (cross) [133]. The experimental values are fitted using an Onsager model. 129

6.17 Photoluminescence efficiency for increasing electric field for pristine Alq$_3$layers (circle) and DCJTB doped Alq$_3$ (cross). The experimental values are fitted using an Onsager model. 129

List of Tables

3.1   Overview of refractive indices, absorption coefficients and charge carrier mobilities 58

3.2   Overview of the electronic and optical material parameters, which have been employed for our simulations. 59

3.3   Overview of rate coefficients for various annihilation processes. Values are given in $(\mathrm{cm}^3/\mathrm{s})$, values for κ_{ISC} are given in $(1/\mathrm{s})$. 61

3.4   Literature values for binding energies of singlet and triplet excitons for various organic semiconductor materials. 68

List of Tables

Bibliography

[1] J. Y. Kim, K. Lee, N. E. Coates, D. Moses, T.-Q. Nguyen, M. Dante, and A. J. Heeger, "Efficient tandem polymer solar cells fabricated by all-solution processing," *Science*, vol. 317, no. 5835, pp. 222–225, 2007.

[2] A. Colsmann, J. Junge, C. Kayser, and U. Lemmer, "Organic tandem solar cells comprising polymer and small-molecule subcells," *Applied Physics Letters*, vol. 89, no. 203506, 2006.

[3] D. Gebeyehu, K. Walzer, G. He, M. Pfeiffer, K. Leo, J. Brandt, A. Gerhardt, and H. Vestweber, "Highly efficient deep-blue organic light emitting diodes with doped transport layers," *Synthetic Metals*, vol. 148, no. 2, pp. 205–211, 2005.

[4] P. J. Delfyett, Ed., *Special issue on High-Efficiency Light-Emitting Diodes*, ser. IEEE Journal of Selected Topics in Quantum Electronics, 2002, vol. 8.

[5] M. Punke, S. Valouch, S. W. Kettlitz, N. Christ, C. Gärtner, M. Gerken, and U. Lemmer, "Dynamic characterization of organic bulk heterojunction photodetectors," *Applied Physics Letters*, vol. 91, no. 071118, 2007.

[6] A. Tsumura, H. Koezuka, and T. Ando, "Macromolecular electronic device: Field-effect transistor with a polythiophene thin film," *Applied Physics Letters*, vol. 49, no. 18, pp. 1210–1212, 1986.

[7] J. Schön, C. Kloc, A. Dodabalapur, and B. Batlogg, "An organic solid state injection laser," *Science*, vol. 289, pp. 599–601, 2000, sR17.

[8] L. Elnadi, L. Alhouty, M. Omar, and M. Ragab, "Organic thin film materials producing novel blue laser," *Chemical Physics Letters*, vol. 286, pp. 9–14, Apr. 1998.

[9] Z. Bao, B. Batlogg, S. Berg, A. Dodabalapur, R. C. Haddon, H. Hwang, C. Kloc, H. Meng, and J. H. Schon, "Retraction," *Science*, vol. 298, no. 5595, pp. 961b–, 2002.

[10] Y. Tian, Z. Gan, Z. Zhou, D. W. Lynch, J. Shinar, J. hun Kang, and Q.-H. Park, "Spectrally narrowed edge emission from organic light-emitting diodes," *Applied Physics Letters*, vol. 91, no. 14, p. 143504, 2007.

[11] H. Nakanotani, C. Adachi, S. Watanabe, and R. Katoh, "Spectrally narrow emission from organic films under continuous-wave excitation," *Applied Physics Letters*, vol. 90, no. 231109, 2007.

[12] N. Tessler, G. Denton, and R. Friend, "Lasing from conjugated-polymer microcavities," *Nature*, vol. 382, pp. 695–697, 1996.

[13] F. Hide, B. Schwartz, M. Diaz-Garcia, and A. Heeger, "Laser emission from solutions and films containing semiconducting polymer and titanium dioxide nanocrystals," *Chemical Physics Letters*, vol. 256, 1996.

[14] I. D. W. Samuel and G. A. Turnbull, "Organic semiconductor lasers," *Chemical Reviews*, vol. 107, pp. 1272–1295, 2007.

[15] A. Rose, Z. Zhu, C. Madigan, T. Swager, and V. Bulovic, "Sensitivity gains in chemosensing by lasing action in organic polymers," *Nature*, vol. 434, pp. 876–879, 2005.

[16] C. Karnutsch, C. Gärtner, V. Haug, U. Lemmer, T. Farrell, B. Nehls, U. Scherf, J. Wang, T. Weimann, G. Heliotis, C. Pflumm, J. deMello, and D. Bradley, "Low threshold blue conjugated polymer lasers with first- and second-order distributed feedback," *Applied Physics Letters*, vol. 89, no. 201108, 2006.

[17] C. Karnutsch, C. Pflumm, G. Heliotis, J. deMello, D. D. C. Bradley, J. Wang, T. Weimann, V. Haug, C. Gärtner, and U. Lemmer, "Improved organic semiconductor lasers based on a mixed-order distributed feedback resonator design," *Applied Physics Letters*, vol. 90, no. 131104, 2007.

[18] M. Stroisch, T. Woggon, U. Lemmer, G. Bastian, and G. Violakis, "Organic semiconductor distributed feedback laser fabricated by direct laser interference ablation," *Optics Express*, pp. 3968–3973, 2006.

[19] C. Karnutsch, V. Haug, C. Gärtner, U. Lemmer, T. Farrell, B. Nehls, U. Scherf, J. Wang, T. Weimann, G. Heliotis, C. Pflumm, J. deMello, and D. Bradley, "Low threshold blue conjugated polymer DFB lasers," *Conference on Lasers and Electro-Optics (CLEO)*, no. CFJ3, 2006.

[20] T. Riedl, T. Rabe, H.-H. Johannes, W. Kowalsky, J. Wang, T. Weimann, P. Hinze, B. Nehls, T. Farrell, and U. Scherf, "Tunable organic thin-film laser pumped by an inorganic violet diode laser," *Applied Physics Letters*, vol. 88, no. 241116, 2006.

[21] A. Vasdekis, G. Tsiminis, J.-C. Ribierre, L. O. Faolain, T. F. Krauss, G. Turnbull, and I. Samuel, "Diode pumped distributed bragg reflector lasers based on a dye-to-polymer energy transfer blend," *Optics Express*, vol. 14, no. 20, pp. 9211–9216, 2006.

[22] C. Karnutsch, M. Stroisch, M. Punke, U. Lemmer, J. Wang, and T. Weimann, "Laser diode pumped organic semiconductor lasers utilizing two-dimensional photonic crystal resonators," *IEEE Photonics Technology Letters*, vol. 19, pp. 741–743, 2007.

[23] Y. Yang, G. A. Turnbull, and I. D. W. Samuel, "Hybrid optoelectronics: A polymer laser pumped by a nitride light-emitting diode," *Applied Physics Letters*, vol. 92, pp. 3306–+, Apr. 2008.

[24] B. Wei, N. Kobayashi, M. Ichikawa, T. Koyama, Y. Taniguchi, and T. Fukuda, "Organic solid laser pumped by an organic light-emitting diode," *Optics Express*, vol. 14, no. 20, pp. 9436– 9443, 2006.

[25] H. Yamamoto, H. Kasajima, W. Yokoyama, H. Sasabe, and C. Adachi, "Extremely-high-density carrier injection and transport over 12000 A/cm^2 into organic thin films," *Applied Physics Letters*, vol. 86, no. 083502, 2005.

[26] N. Tessler, N. Harrison, and R. Friend, "High peak brightness polymer light-emitting diodes," *Advanced Materials*, vol. 10, pp. 64–68, 1998.

[27] N. Tessler, "Lasers based on semiconducting organic materials," *Advanced Materials*, vol. 11, pp. 363–370, 1999.

[28] I. Campbell, D. Smith, C. Neef, and J. Ferraris, "Charge transport in polymer light-emitting diodes at high current density," *Applied Physics Letters*, vol. 75, no. 6, pp. 841–843, 1999.

[29] W. Yokoyama, H. Sasabe, and C. Adachi, "Carrier injection and transport of steady-state high current density exceeding 1000 A/cm^2 in organic thin films," *Jpn. J. Appl. Phys.*, vol. 42, pp. L1353–L1355, 2003.

[30] H. Nakanotani, H. Sasabe, and C. Adachi, "Low lasing threshold in organic distributed feedback solid state lasers using bisstyrylbenzene derivative as active material," in *Proceedings of SPIE*, ser. Organic Light-Emitting Materials and Devices IX, vol. 5937, no. 59370W, San Diego, CA, USA, 2005.

[31] A. Heeger, "Nobel lecture: Semiconducting and metallic polymers: The fourth generation of polymeric materials," *Reviews of Modern Physics*, vol. 73, pp. 681–700, 2001.

[32] J. Burroughes, D. Bradley, A. Brown, R. Marks, K. Mackay, R. Friend, P. Burn, and A. Holmes, "Light-emitting diodes based on conjugated polymers," *Nature*, vol. 347, pp. 539–541, 1990.

[33] F. Hide, M. Diaz-Garcia, B. Schwartz, M. Andersson, and A. Heeger, "Semiconducting polymers: A new class of solid-state laser materials," *Science*, vol. 273, pp. 1833–, 1996.

[34] V. Kozlov, V. Bulovic, P. Burrows, and S. Forrest, "Laser action in organic semiconductor waveguide and double-heterostructure devices," *Nature*, vol. 389, pp. 362–364, 1997.

[35] M. Berggren, A. Dodabalapur, R. Slusher, and Z. Bao, "Light amplification in organic thin films using cascade energy transfer," *Nature*, vol. 389, pp. 466–469, 1997.

[36] P. Brocorens, E. Zojer, and J. Cornil, "Theoretical characterization of phenylene-based oligomers, polymers, and dendrimers," *Synthetic Metals*, vol. 100, pp. 141–162, 1999.

[37] J. M. Lupton, "Nanoengineering of organic light-emitting diodes," Dissertation, University of Durham, 2000.

[38] A. Otomo, S. Yokoyama, T. Nakahama, and S. Mashiko, "Supernarrowing mirrorless laser emission in dendrimer-doped polymer waveguides," *Applied Physics Letters*, vol. 77, no. 24, pp. 3881–3883, 2000.

[39] J. Lupton, I. Samuel, R. Beavington, P. Burn, and H. Bässler, "Control of mobility in molecular organic semiconductors by dendrimer generation," *Physical Review B*, vol. 63, p. 155206, 2001.

[40] J. Lupton, I. Samuel, and P. Burn, "Origin of spectral broadening in p-conjugated amorphous semiconductors," *Physical Review B*, vol. 155206, pp. 1–6, 2002.

[41] J. Markham, S. Lo, S. Magennis, P. Burn, and I. Samuel, "High-efficiency green phosphorescence from spin-coated single-layer dendrimer light-emitting diodes," *Applied Physics Letters*, vol. 80, no. 15, pp. 2645–2647, 2002.

[42] I. Samuel and P. Burn, "Light-emitting dendrimers," *Abstract*, 2003, online-Presentation.

[43] J. Lawrence, G. Turnbull, I. Samuel, G. Richards, and P. Burn, "Optical amplification in a first-generation dendritic organic semiconductor," *Optics Letters*, vol. 29, no. 8, pp. 869–871, 2004.

[44] J. Markham, I. Samuel, S. Lo, P. Burn, M. Weiter, and H. Bässler, "Charge transport in highly efficient iridium cored electrophosphorescent dendrimers," *Journal of Applied Physics*, vol. 95, no. 2, pp. 438–445, 2004.

[45] P. A. Levermore, R. Xia, W. Lai, X. H. Wang, W. Huang, and D. D. C. Bradley, "Deep-blue light emitting triazatruxene core/oligo-fluorene branch dendrimers for electroluminescence and optical gain applications," *Journal of Physics D: Applied Physics*, vol. 40, no. 7, pp. 1896–1901, 2007.

[46] S.-C. Lo and P. Burn, "Development of dendrimers: Macromolecules for use in organic light-emitting diodes and solar cells," *Chemical Reviews*, vol. 107, pp. 1097–1116, 2007.

[47] H. Haken and H. Wolf, *Molekülphysik und Quantenchemie*. Berlin: Springer Verlag, 1998.

[48] P. Prins, F. Grozema, B. Nehls, T. Farrell, U. Scherf, and L. Siebbeles, "Enhanced charge-carrier mobility in b-phase polyfluorene," *Physical Review B*, vol. 74, p. 113203, 2006.

[49] H. Bässler, G. Schönherr, M. Abkowitz, and D. Pai, "Hopping transport in prototypical organic glasses," *Physical Review B*, vol. 26, pp. 3105–3113, 1982.

[50] M. Reufer, S. Riechel, J. Lupton, J. Feldmann, D. Schneider, T. Benstem, T. Dobbertin, W. Kowalsky, U. Scherf, K. Forberich, A. Gombert, V. Wittwer, and U. Lemmer, "Low-threshold polymeric distributed feedback lasers with metallic contacts," *Applied Physics Letters*, vol. 84, no. 17, pp. 3262–3264, 2004.

[51] H. Bässler, V. Arkhipov, E. Emelianova, A. Gerhard, A. Hayer, C. Im, and J. Rissler, "Excitons in p-conjugated polymers," *Synthetic Metals*, vol. 135-136, pp. 377–382, 2003.

[52] M. Pope and C. Swenberg, "Electronic processes in organic crystals," 1982.

[53] M. Reufer, M. Walter, P. Lagoudakis, A. Hummel, J. Kolb, H. Roskos, U. Scherf, and J. Lupton, "Spin-conserving carrier recombination in conjugated polymers," *Nature Materials*, vol. 4, pp. 1–7, 2005.

[54] J. Scott and G. Malliaras, "Charge injection and recombination at the metal-organic interface," *Chemical Physics Letters*, vol. 299, pp. 115–119, 1999.

[55] S. M. Sze, *Physics of semiconductor devices*, 1981.

[56] J. G. Simmons, "Richardson-schottky effect in solids," *Phys. Rev. Lett.*, vol. 15, no. 25, pp. 967–968, Dec 1965.

[57] P. R. Emtage and J. J. O'Dwyer, "Richardson-schottky effect in insulators," *Phys. Rev. Lett.*, vol. 16, no. 9, pp. 356–358, Feb 1966.

[58] V. Arkhipov, E. Emelianova, Y. Tak, and H. Bässler, "Charge injection into light-emitting diodes: Theory and experiment," *Journal of Applied Physics*, vol. 84, no. 2, pp. 848–, 1998.

[59] T. Förster, "Zwischenmolekulare energiewanderung und fluoreszenz," *Annalen der Physik*, vol. 6, pp. 55–75, 1948.

[60] D. L. Dexter, "A theory of sensitized luminescence in solids," *The Journal of Chemical Physics*, vol. 21, no. 5, pp. 836–850, 1953.

[61] M. Baldo, D. O'Brien, M. Thompson, and S. Forrest, "Excitonic singlet-triplet ratio in a semiconducting organic thin film," *Physical Review B*, vol. 60, no. 20, pp. 14 422–14 428, 1999.

[62] H. Nakanotani, H. Sasabe, and C. Adachia, "Singlet-singlet and singlet-heat annihilations in fluorescence-based organic light-emitting diodes under steady-state high current density," *Applied Physics Letters*, vol. 86, no. 213506, 2005.

[63] I. Sokolik, R. Priestley, A. Walser, R. Dorsinville, and C. Tang, "Bimolecular reactions of singlet excitons in tris(8-hydroxyquinoline) aluminum," *Applied Physics Letters*, vol. 69, no. 27, pp. 4168–4170, 1996.

[64] H. Nakanotani, T. Oyamada, Y. Kawamura, H. Sasabe, and C. Adachi, "Injection and transport of high current density over 1000 A/cm^2 in organic light emitting diodes under pulse excitation," *Japanese Journal of Applied Physics*, vol. 44, no. 6A, pp. 3659–3662, 2005.

[65] S. Lim, T. G. Bjorklund, and C. Bardeen, "The role of long-lived dark states in the photoluminescence dynamics of poly.phenylene vinylene. conjugated polymers. ii. excited-state quenching versus ground-state depletion," *Journal of Chemical Physics*, vol. 118, no. 9, pp. 4297–4305, 2003.

[66] D. Hertel and K. Meerholz, "Triplet-polaron quenching in conjugated polymers," *Journal of Physical Chemistry B*, vol. 111, no. 42, pp. 12 075–12 080, 2007.

[67] P. Lane, X. Wei, and Z. Vardeny, "Studies of charged excitations in p-conjugated oligomers and polymers by optical modulation," *Physical Review Letters*, vol. 77, no. 8, pp. 1544–1546, 1996.

[68] R. H. Friend, D. D. C. Bradley, and P. D. Townsend, "Photoexcitation in conjugated polymers," *Journal of Physics D-Applied Physics*, vol. 20, no. 11, pp. 1367–1384, 1987.

[69] Z. Vardeny, E. Ehrenfreund, O. Brafman, M. Nowak, H. Schaffer, A. J. Heeger, and F. Wudl, "Photogeneration of confined soliton pairs (bipolarons) in poly-thiophene," *Phys. Rev. Lett.*, vol. 56, no. 6, pp. 671–674, Feb 1986.

[70] T.-C. Chung, J. H. Kaufman, A. J. Heeger, and F. Wudl, "Charge storage in doped poly(thiophene): Optical and electrochemical studies," *Phys. Rev. B*, vol. 30, no. 2, pp. 702–710, Jul 1984.

[71] K. E. Ziemelis, A. T. Hussain, D. D. C. Bradley, R. H. Friend, J. Rühe, and G. Wegner, "Optical spectroscopy of field-induced charge in poly(3-hexyl thienylene) metal-insulator-semiconductor structures: Evidence for polarons," *Phys. Rev. Lett.*, vol. 66, no. 17, pp. 2231–2234, Apr 1991.

[72] D. Fichou, G. Horowitz, B. Xu, and F. Garnier, "Stoichiometric control of the successive generation of the radical cation and dication of extended [alpha]-conjugated oligothiophenes: a quantitative model for doped polythiophene," *Synthetic Metals*, vol. 29, pp. 243–259, 1990.

[73] M. Deussen and H. Bassler, "Anion and cation absorption spectra of conjugated oligomers and polymers," *Chemical Physics*, vol. 164, pp. 247–257, 1992.

[74] J. Cornil and J.-L. Bredas, "Nature of the optical transitions in charged oligothiophenes," *Advanced Materials*, vol. 7, pp. 295–297, 1995.

[75] Y. Shimoi, S. Abe, and K. Harigaya, "Theory of optical absorption in doped conjugated polymers," *Molecular Crystals and Liquid Crystals*, vol. 267, pp. 329–334, 1995.

[76] J.-y. Fu, J.-f. Ren, D.-s. Liu, and S.-j. Xie, "Polarons and bipolarons in polythiophene contacted with half-metallic ferromagnetic manganite," *Thin Solid Films*, vol. 503, pp. 196–200, 2006.

[77] S. Frolov, W. Gellermann, Z. Vardeny, M. Ozaki, and K. Yoshino, "Picosecond photophysics of luminescent conducting polimers; from excitons to polaron pairs." *Synthetic Metals*, vol. 84, pp. 493–496, 1997.

[78] S.-C. Yang, W. Graupner, S. Guha, P. Puschnig, C. Martin, H. Chandrasekhar, M. Chandrasekhar, G. Leising, C. Ambrosch-Draxl, and U. Scherf, "Geometry-dependent electronic properties of highly fluorescent conjugated molecules," *Physical Review Letters*, vol. 85, no. 11, pp. 2388–2391, 2000.

[79] C. Belton, D. O'Brien, W. Blau, A. Cadby, P. Lane, D. Bradley, H. Byrne, R. Stockmann, and H. Hörbold, "Excited-state quenching of a highly luminescent conjugated polymer," *Applied Physics Letters*, vol. 78, no. 8, pp. 1059–1061, 2001.

[80] P. Brown, H. Sirringhaus, M. Harrison, M. Shkunov, and R. Friend, "Optical spectroscopy of field-induced charge in self-organized high mobility poly(3-hexylthiophene)," *Physical Review B*, vol. 63, pp. 125 204–1–11, 2001.

[81] I. Campbell, D. Smith, C. Neef, and J. Ferraris, "Optical properties of single carrier polymer diodes under high electrical injection," *Applied Physics Letters*, vol. 78, no. 3, pp. 270–272, 2001.

[82] I. Campbell, D. L. Smith, C. Neef, and J. Ferraris, "Charge-carrier effects on the optical properties of poly.p-phenylene vinylene," *Physical Review B*, vol. 64, pp. 035 203/1–035 203/5, 2001.

[83] C. Silva, D. Russell, A. Dhoot, L. Herz, C. Daniel, N. Greenham, A. Arias, S. Setayesh, K. Müllen, and R. Friend, "Exciton and polaron dynamics in a step-ladder polymeric semiconductor: the influence of interchain order," *Journal of Physics: Condensed Matter*, vol. 14, pp. 9803–9824, 2002.

[84] K. Book, H. Bässler, A. Elschner, and S. Kirchmeyer, "Hole injection from an itojpedt anode into the hole transporting layer of an oled probed by bias induced absorption," *Organic Electronics*, vol. 4, pp. 227–232, 2003.

[85] R. Österbacka, M. Wohlgenannt, M. Shkunov, D. Chinn, and Z. Vardeny, "Excitons, polarons, and laser action in poly.(p-phenylene vinylene) films," *Journal of Chemical Physics*, vol. 118, no. 19, pp. 8905–8916, 2003.

[86] C. Pflumm, C. Karnutsch, M. Gerken, and U. Lemmer, "Gain and polaron absorption in electrically pumped single-layer organic laser diodes," in *Proceedings of 34th European Solid-State Device Research Conference (ESSDERC)*, Leuven, Belgium, 2004.

[87] M. Wohlgenannt, "Polarons in p-conjugated semiconductors: absorption spectroscopy and spin-dependent recombination," *Physica Status Solidi A*, vol. 201, no. 6, pp. 1188–1204, 2004.

[88] A. Ye, Z. Shuai, O. Kwon, J.-L. Bredas, and D. Beljonne, "Optical properties of singly charged conjugated oligomers: A coupled-cluster equation of motion study," *Journal of Chemical Physics*, vol. 121, no. 12, pp. 5567–5578, 2004.

[89] S. Lattante, F. Romano, A. Caricato, M. Martino, and M. Anni, "Low electrode induced optical losses in organic active single layer polyfluorene waveguides with two indium tin oxide electrodes deposited by pulsed laser deposition," *Applied Physics Letters*, vol. 89, no. 031108.

[90] M. Harrison, K. Ziemelis, R. Friend, P. Burn, and A. Holmes, "Optical spectroscopy of field-induced charge in poly(2,5-dimethoxy-p-phenylene vinylene) metal-insulator-semiconductor structures," *Synthetic Metals*, vol. 55-57, pp. 218–223, 1993.

[91] V. Kozlov, P. Burrows, G. Parthasarathy, and S. Forrest, "Optical properties of molecular organic semiconductor thin films under intense electrical excitation," *Applied Physics Letters*, vol. 74, no. 8, pp. 1057–1059, 1999.

[92] G. Lanzani, S. Frolov, M. Nisoli, P. Lane, S. De Silvestri, R. Tubino, F. Abbate, and Z. Vardeny, "Ultrafast spectroscopy of photoexcitations in α-sexithienyl films: evidence for excitons and polaron-pairs," *Synthetic Metals*, vol. 84, pp. 517–520, 1997.

[93] A. Dhoot, D. Ginger, D. Beljonne, Z. Shuai, and N. Greenham, "Triplet formation and decay in conjugated polymer devices," *Chemical Physics Letters*, vol. 360, pp. 195–201, 2002.

[94] S. Riechel, "Organic semiconductor lasers with two-dimensional distributed feedback," Dissertation, Universität München, Dissertation, 2002.

[95] G. Barlow and K. Shore, "Threshold current analysis of distributed feed-back organic semiconductor lasers," *IEE Proceedings Optoelectronics*, vol. 148, no. 1, pp. 2–6, 2001.

[96] R. Xia, G. Heliotis, and D. Bradley, "Semiconducting polyfluorenes as materials for solid-state polymer lasers across the visible spectrum," *Synthetic Metals*, vol. 140, pp. 117–120, 2004.

[97] G. Heliotis, D. Bradley, M. Goossens, S. Richardson, G. Turnbull, and I. Samuel, "Operating characteristics of a traveling-wave semiconducting polymer optical amplifier," *Applied Physics Letters*, vol. 85, no. 25, pp. 6122–6124, 2004.

[98] D. Pisignano, M. Anni, G. Gigli, R. Cingolani, M. Zavelani-Rossi, G. Lanzani, G. Barbarella, and L. Favaretto, "Amplified spontaneous emission and efficient tunable laser emission from a substituted thiophene-based oligomer," *Applied Physics Letters*, vol. 81, no. 19, pp. 3534–3536, 2002.

[99] W. Holzer, A. Penzkofer, S.-H. Gong, D. Bradley, X. Long, and A. Bleyer, "Effective stimulated emission and excited-state absorption cross-section spectra of poly(m-phenylenevinylene-co-2,5-dioctoxy-p-phenylenevinylene) and t,t'-didecyloxy-ii-distyrylbenzene," *Chemical Physics*, vol. 224, no. 2, pp. 315–326, 1997.

[100] L. Candeias, G. Padmanaban, and S. Ramakrishnan, "The effect of broken conjugation on the optical absorption spectra of the triplet states of isolated chains of poly(phenylene vinylene)s," *Chemical Physics Letters*, vol. 349, pp. 394–398, 2001.

[101] A. P. Monkman, H. D. Burrows, M. da G. Miguel, I. Hamblett, and S. Navaratnam, "Measurement of the $S_0 - T_1$ energy gap in poly(2-methoxy,5-(2'-ethyl-hexoxy)-p-phenylenevinylene) by triplet-triplet energy transfer," *Chemical Physics Letters*, vol. 307, no. 5-6, pp. 303–309, 1999.

[102] V. Cleave, G. Yahioglu, P. Le Barny, D. H. Hwang, A. B. Holmes, R. H. Friend, and N. Tessler, "Transfer processes in semiconducting polymer-porphyrin blends," *Advanced Materials*, vol. 13, no. 1, pp. 44–47, 2001.

[103] Z. Shuai and J. L. Brédas, "Coupled-cluster approach for studying the electronic and nonlinear optical properties of conjugated molecules," *Physical Review B*, vol. 62, no. 23, p. 15452, 2000.

[104] R. Bartlett, "Coupled-cluster approach to molecular structure and spectra: a step toward predictive quantum chemistry," *Journal of Physical Chemistry*, vol. 93, p. 1697, 1989.

[105] C.-L. Lee, X. Yang, and N. C. Greenham, "Determination of the triplet excited-state absorption cross section in a polyfluorene by energy transfer from a phosphorescent metal complex," *Physical Review B (Condensed Matter and Materials Physics)*, vol. 76, no. 24, p. 245201, 2007.

[106] C. Rothe, S. King, and A. Monkman, "Electric-field-induced singlet and triplet exciton quenching in films of the conjugated polymer polyspirobifluorene," *Physical Review B*, vol. 72, p. 085220, 2005.

[107] J. Kalinowski, W. Stampor, J. Mezyk, M. Cocchi, D. Virgili, V. Fattori, and P. D. Marco, "Quenching effects in organic electrophosphorescence," *Physical Review B*, vol. 66, no. 235321, 2002.

[108] S. F. Alvarado, P. F. Seidler, D. G. Lidzey, and D. D. C. Bradley, "Direct determination of the exciton binding energy of conjugated polymers using a scanning tunneling microscope," *Physical Review Letters*, vol. 81, no. 5, pp. 1082–1085, 1998.

[109] T. G. Pedersen, T. B. Lynge, P. K. Kristensen, and P. M. Johansen, "Theoretical study of conjugated porphyrin polymers," *Thin Solid Films*, vol. 477, no. 1-2, pp. 182–186, 2005.

[110] M. Knupfer, "Exciton binding energies in organic semiconductors," *Applied Physics a-Materials Science and Processing*, vol. 77, no. 5, pp. 623–626, 2003.

[111] K. Hummer, P. Puschnig, and C. Ambrosch-Draxl, "Lowest optical excitations in molecular crystals: Bound excitons versus free electron-hole pairs in anthracene," *Physical Review Letters*, vol. 92, no. 14, p. 147402, 2004.

[112] I. Hill, A. Kahn, Z. Soos, and R. Pascal (Jr.), "Charge-separation energy in films of p-conjugated organic molecules," *Chemical Physics Letters*, vol. 327, pp. 181–188, 2000.

[113] M. Knupfer, H. Peisert, and T. Schwieger, "Band-gap and correlation effects in the organic semiconductor Alq$_3$," *Phys. Rev. B*, vol. 65, no. 033204, Dec 2001.

[114] B. Schweitzer and H. Bässler, "Excitons in conjugated polymers," *Synthetic Metals*, vol. 109, no. 1-6, pp. 1–6, 2000.

[115] E. List, C. Kim, A. Naik, U. Scherf, G. Leising, W. Graupner, and J. Shinar, "Interaction of singlet excitons with polarons in wide band-gap organic semiconductors: A quantitative study," *Physical Review B*, vol. 64, no. 155204, 2001.

[116] G. Weiser, "Stark effect of one-dimensional wannier excitons in polydiacetylene single crystals," *Physical Review B*, vol. 45, no. 24, pp. 14076–14085, 1992.

[117] M. Scheidler, U. Lemmer, R. Kersting, S. Karg, W. Riess, B. Cleve, R. F. Mahrt, H. Kurz, H. Bässler, E. O. Göbel, and P. Thomas, "Monte carlo study of picosecond exciton relaxation and dissociation in poly(phenylenevinylene)," *Phys. Rev. B*, vol. 54, no. 8, pp. 5536–5544, Aug 1996.

[118] A. Haugeneder, M. Neges, C. Kallinger, W. Spirkl, U. Lemmer, J. Feldmann, U. Scherf, E. Harth, A. Gügel, and K. Müllen, "Exciton diffusion and dissociation in conjugated polymer/fullerene blends and heterostructures," *Physical Review B*, vol. 59, no. 23, pp. 15 346 – 15 351, 1999.

[119] C. I. Wu, Y. Hirose, H. Sirringhaus, and A. Kahn, "Electron-hole interaction energy in the organic molecular semiconductor PTCDA," *Chemical Physics Letters*, vol. 272, pp. 43–47, 1997.

[120] L. Onsager, "Initial recombination of ions," *Phys. Rev.*, vol. 54, no. 8, pp. 554–557, 1938.

[121] C. L. Braun, "Electric field assisted dissociation of charge transfer states as a mechanism of photocarrier production," *Journal of Chemical Physics*, vol. 80, pp. 4157–4161, 1984.

[122] T. E. Goliber and J. H. Perlstein, "Analysis of photogeneration in a doped polymer system in terms of a kinetic model for electric-field-assisted dissociation of charge-transfer states," *Journal of Chemical Physics*, vol. 80, pp. 4162–4167, 1984.

[123] H. Najafov, I. Biaggio, T.-K. Chuang, and M. Hatalis, "Exciton dissociation by a static electric field followed by nanoscale charge transport in ppv polymer films," *Physical Review B*, vol. 73, no. 125202, 2006.

[124] J. Kalinowski, W. Stampor, J. Szmytkowski, D. Virgili, M. Cocchi, V. Fattori, and C. Sabatini, "Coexistence of dissociation and annihilation of excitons on charge carriers in organic phosphorescent emitters," *Physical Review B*, vol. 74, no. 085316, 2006.

[125] J. Szmytkowski, W. Stampor, J. Kalinowski, and Z. Kafafi, "Electric field-assisted dissociation of singlet excitons in tris-(8-hydroxyquinolinato) aluminum (iii)," *Applied Physics Letters*, vol. 80, no. 8, pp. 1465–1467, 2002.

[126] J. Kalinowski, W. Stampor, and P. D. Marco, "Electromodulation of luminescence in organic photoconductors," *Journal of The Electrochemical Society*, vol. 143, no. 1, pp. 315–325, 1996.

[127] M. Gailberger and H. Bässler, "dc and transient photoconductivity of poly(2-phenyl-1,4-phenylenevinylene)," *Physical Review B*, vol. 44, no. 16, pp. 8643–8651, Oct 1991.

[128] R. Kersting, U. Lemmer, M. Deussen, H. Bakker, R. Mahrt, H. Kurz, V. I. Arkhipov, H. Bässler, and E. Göbel, "Ultrafast field-induced dissiciation of excitons in conjugated polymers," *Physical Review Letters*, vol. 73, no. 10, pp. 1440–1443, 1994.

[129] R. Kersting, U. Lemmer, R. Mahrt, K. Leo, H. Kurz, H. Bässler, and E. Göbel, "Femtosecond energy relaxation in p-conjugated polymers," *Physical Review Letters*, vol. 70, no. 24, pp. 3820–3823, 1993.

[130] W. Stampor, J. Kalinowski, P. DiMarco, and V. Fattori, "Electric field effect on luminescence efficiency in 8-hydroxyquinoline aluminum (alq3) thin films," *Applied Physics Letters*, vol. 70, no. 15, pp. 1935–1937, 1997.

[131] M. Deussen, M. Scheidler, and H. Bässler, "Electric field-induced photoluminescence quenching in thin-film lightemitting diodes based on poly(phenyl-p-phenylene vinylene)," *Synthetic Metals*, vol. 73, pp. 123–129, 1995.

[132] Y. Luo, H. Aziz, G. Xu, and Z. Popovic, "Electron-induced quenching of excitons in luminescent materials," *Chemistry of Materials*, vol. 19, no. 9, pp. 2288–2291, 2007.

[133] Y. Luo, H. Aziz, Z. D. Popovic, and G. Xu, "Electric-field-induced fluorescence quenching in dye-doped tris(8-hydroxyquinoline) aluminum layers," *Applied Physics Letters*, vol. 89, no. 10, p. 103505, 2006.

[134] T. H. Maiman, "Stimulated optical radiation in ruby," *Nature*, vol. 187, no. 4736, pp. 493–494, 1960.

[135] B. Soffer and B. McFarland, "Continuously tunable, narrow-band organic dye lasers," *Applied Physics Letters*, vol. 10, no. 10, pp. 266–267, 1967.

[136] N. Karl, "Laser emission from an organic molecular crystal," *Physica Status Solidi A*, vol. 13, p. 651, 1972.

[137] D. Moses, "High quantum efficiency luminescence from a conducting polymer in solution: A novel polymer laser dye," *Applied Physics Letters*, vol. 60, no. 26, pp. 3215–3216, 1992.

[138] C. Kallinger, M. Hilmer, and A. Haugeneder, "A flexible conjugated polymer laser," *Advanced Materials*, vol. 10, no. 12, pp. 920–923, 1998, sR23.

[139] C. Karnutsch, V. Haug, C. Gärtner, U. Lemmer, T. Farrell, B. Nehls, U. Scherf, J. Wang, T. Weimann, G. Heliotis, C. Pflumm, J. deMello, and D. Bradley, "Low threshold blue conjugated polymer dfb lasers," *Conference on Lasers and Optoelectronics (CLEO), Long Beach, CA, USA*, p. Paper No. CFJ3, 2006.

[140] F. De Martini and J. Jacobowitz, "Anomalous spontaneous-stimulated-decay phase transition and zero-threshold laser action in a microscopic cavity," *Physical Review Letters*, vol. 60, no. 17, pp. 1711 – 1714, 1988.

[141] M. Diaz-Garcia, F. Hide, B. Schwartz, M. Andersson, Q. Pei, and A. Heeger, "Plastic lasers: Semiconducting polymers as a new class of solid-state laser materials," *Synthetic Metals*, vol. 84, pp. 455–462, 1997.

[142] S. E. Burns, G. Denton, N. Tessler, M. A. Stevens, F. Cacialli, and R. H. Friend, "High finesse organic microcavities," *Optical Materials*, vol. 9, no. 1-4, pp. 18–24, 1998.

[143] A. Dodabalapur, M. Berggren, and R. Slusher, "Resonators and materials for organic lasers based on energy transfer," *IEEE Journal of Selected Topics in Quantum Electronics*, vol. 4, no. 1, pp. 67–74, 1998, sR28.

[144] V. Kozlov, V. Bulovic, P. Burrows, M. Baldo, V. Khalfin, G. Parthasarathy, S. Forrest, Y. You, and M. Thompson, "Study of lasing action based on förster energy transfer in optically pumped organic semiconductor thin films," *Journal of Applied Physics*, vol. 84, no. 8, pp. 4096–4108, 1998.

[145] V. Kozlov, V. Bulovic, and S. Forrest, "Temperature independent performance of organic semiconductor lasers," *Applied Physics Letters*, vol. 71, no. 18, pp. 2575–2577, 1997.

[146] M. Ichikawa, R. Hibino, M. Inoue, T. Haritani, S. Hotta, K. Araki, T. Koyama, and Y. Taniguchi, "Laser oscillation in monolithic molecular single crystals," *Advanced Materials*, vol. 17, pp. 2073–2077, 2005.

[147] X. Zhu, D. Gindre, N. Mercier, P. Frère, and J. M. Nunzi, "Stimulated emission from a needle-like single crystal of an end-capped fluorene/phenylene co-oligomer," *Advanced Materials*, vol. 15, no. 11, pp. 906–909, 2003.

[148] M. Ichikawa, K. Nakamura, M. Inoue, H. Mishima, T. Haritani, R. Hibino, T. Koyama, and Y. Taniguchi, "Photopumped laser oscillation and charge-injected luminescence from organic semiconductor single crystals of a thiophene/phenylene co-oligomer," *Applied Physics Letters*, vol. 87, pp. 221 113 1–221 113-3, 2005.

[149] A. Schülzgen, C. Spiegelberg, M. M. Morrell, S. B. Mendes, P. M. Allemand, Y. Kawabe, M. Kuwata-Gonokami, S. Honkanen, M. Fallahi, B. Kippelen, and N. Peyghambarian, "Light amplification and laser emission in conjugated polymers," *Optical Engineering*, vol. 37, no. 4, pp. 1149–1156, 1998.

[150] Y. Kawabe, C. Spiegelberg, A. Schulzgen, M. Nabor, B. Kippelen, E. Mash, P. Allemand, M. Kuwata-Gonokami, K. Takeda, and N. Peyghambarian, "Whispering-gallery-mode microring laser using a conjugated polymer," *Applied Physics Letters*, vol. 72, no. 2, pp. 141–143, 1998.

[151] S. Frolov, Z. Vardeny, and K. Yoshino, "Plastic microring lasers on fibers and wires," *Applied Physics Letters*, vol. 72, no. 15, pp. 1802–1804, 1998.

[152] S. Frolov, A. Fujii, D. Chinn, M. Hirohata, R. Hidayat, M. Taraguchi, T. Masuda, K. Yoshino, and Z. Vardeny, "Microlasers and micro-leds from disubstituted polyacetylene," *Advanced Materials*, vol. 10, no. 11, pp. 869–872, 1998.

[153] M. Berggren, A. Dodabalapur, Z. Bao, and R. Slusher, "Solid-state droplet laser made from an organic blend with a conjugated polymer emitter," *Advanced Materials*, vol. 9, no. 12, pp. 968–971, 1997.

[154] G. Turnbull, T. Krauss, W. Barnes, and I. Samuel, "Tuneable distributed feedback lasing in meh-ppv films," *Synthetic Metals*, vol. 121, pp. 1757–1758, 2001.

[155] G. Turnbull, A. Carleton, A. Tahraouhi, T. F. Krauss, I. Samuel, G. Barlow, and K. A. Shore, "Effect of gain localization in circular-grating distributed feedback lasers," *Applied Physics Letters*, vol. 87, pp. 201 101–1 –201 101–3, 2005.

[156] G. Heliotis, R. Xia, D. Bradley, G. Turnbull, I. Samuel, P. Andrew, and W. Barnes, "Two-dimensional distributed feedback lasers using a broadband, red polyfluorene gain medium," *Journal of Applied Physics*, vol. 96, no. 12, pp. 6959–6965, 2004.

[157] G. Heliotis, S. Choulis, G. Itskos, R. Xia, R. Murray, P. N. Stavrinou, and D. Bradley, "Low-threshold lasers based on a high-mobility semiconducting polymer," *Applied Physics Letters*, vol. 88, no. 081104, 2006.

[158] S. Riechel, U. Lemmer, J. Feldmann, S. Berleb, A. Mückl, W. Brütting, A. Gombert, and V. Wittwer, "Very compact tunable solid-state laser utilizing a thin-film organic semiconductor," *Optics Letters*, vol. 26, no. 9, pp. 593–595, 2001.

[159] D. Schneider, S. Hartmann, T. Bernstem, T. Dobbertin, D. Heithecker, D. Metzdorf, E. Becker, T. Riedl, H. Johannes, W. Kowalsky, T. Weimann, J. Wang, and P. Hinze, "Wavelength-tunable organic solid-state distributed feedback laser," *Applied Physics B*, vol. 77, pp. 399–402, 2003.

[160] M. Weinberger, G. Langer, A. Pogantsch, A. Haase, E. Zojer, and W. Kern, "Continuously color-tunable rubber laser," *Advanced Materials*, vol. 16, no. 2, pp. 130–133, 2004.

[161] J. Lawrence, Y. Ying, P. Jiang, and S. Foulger, "Dynamic tuning of organic lasers with colloidal crystals," *Advanced Materials*, vol. 18, pp. 300–303, 2006.

[162] S. Schols, S. Verlaak, C. Rolin, D. Cheyns, J. Genoe, and P. Heremans, "An organic light-emitting diode with field-effect electron transport," *Advanced Functional Materials*, vol. 18, pp. 136– 144, 2008.

[163] T. Matsushima, H. Sasabe, and C. Adachi, *Applied Physics Letters*, vol. 88, no. 033508, 2006.

[164] C. Pflumm, C. Karnutsch, R. Boschert, M. Gerken, U. Lemmer, J. deMello, and D. Bradley, "Modelling of the laser dynamics of electrically pumped organic semiconductor laser diodes," in *Proceedings of SPIE*, ser. Organic Light-Emitting Materials and Devices IX, vol. 5937, no. 59370X, San Diego, CA, USA, 2005.

[165] M. Reufer, J. Feldmann, P. Rudati, A. Ruhl, D. Müller, K. Meerholz, C. Karnutsch, M. Gerken, and U. Lemmer, "Amplified spontaneous emission in an organic semiconductor multilayer waveguide structure including a highly conductive transparent electrode," *Applied Physics Letters*, vol. 86, no. 221102, 2005.

[166] C. Pflumm, C. Karnutsch, M. Gerken, and U. Lemmer, "Parametric study of modal gain and threshold power density in electrically pumped single-layer organic optical amplifier and laser diode structures," *IEEE Journal of Quantum Electronics*, vol. 41, no. 3, pp. 316–336, 2005.

[167] P. Görrn, T. Rabe, T. Riedl, and W. Kowalsky, "Loss reduction in fully contacted organic laser waveguides using TE_2 modes," *Applied Physics Letters*, vol. 91, no. 041113, 2007.

[168] P. Görrn, T. Rabe, T. Riedl, W. Kowalsky, F. Galbrecht, and U. Scherf, "Low loss contacts for organic semiconductor lasers," *Applied Physics Letters*, vol. 89, no. 161113, 2006.

[169] B. Yoo, T. Jung, D. Basu, A. Dodabalapur, B. A. Jones, A. Facchetti, M. R. Wasielewski, and T. J. Marks, "High-mobility bottom-contact n-channel organic transistors and their use in complementary ring oscillators," *Applied Physics Letters*, vol. 88, pp. 2104–+, Feb. 2006.

[170] A. Dodabalapur, H. E. Katz, L. Torsi, and R. C. Haddon, "Organic Heterostructure Field-Effect Transistors," *Science*, vol. 269, pp. 1560–1562, Sep. 1995.

[171] T. D. Anthopoulos, B. Singh, N. Marjanovic, N. S. Sariciftci, A. Montaigne Ramil, H. Sitter, M. Cölle, and D. M. de Leeuw, "High performance n-channel organic field-effect transistors and ring oscillators based on C_{60} fullerene films," *Applied Physics Letters*, vol. 89, pp. 3504–+, Nov. 2006.

[172] J. H. Na, M. Kitamura, and Y. Arakawa, "High performance n-channel thin-film transistors with an amorphous phase C_{60} film on plastic substrate," *Applied Physics Letters*, vol. 91, pp. 3501–+, Nov. 2007.

[173] H. Sirringhaus, N. Tessler, and R. H. Friend, "Integrated Optoelectronic Devices Based on Conjugated Polymers," *Science*, vol. 280, pp. 1741–+, Jun. 1998.

[174] W. Geens, S. E. Shaheen, B. Wessling, C. J. Brabec, J. Poortmans, and N. Serdar Sariciftci, "Dependence of field-effect hole mobility of ppv-based polymer films on the spin-casting solvent," *Organic Electronics*, vol. 3, no. 3-4, pp. 105–110, 2002, tY - JOUR.

[175] T. Matsushima, K. Fujita, and T. Tsutsui, "High Field-Effect Hole Mobility in Organic-Inorganic Hybrid Thin Films Prepared by Vacuum Vapor Deposition Technique," *Japanese Journal of Applied Physics*, vol. 43, pp. 1199–+, Sep. 2004.

[176] C. Gärtner, C. Karnutsch, U. Lemmer, and C. Pflumm, "The influence of annihilation processes on the threshold current density of organic laser diodes," *Journal of Applied Physics*, vol. 101, p. 023107, 2007.

[177] C. Gärtner, C. Karnutsch, C. Pflumm, and U. Lemmer, "Numerical device simulation of double heterostructure organic laser diodes including current induced absorption processes," *IEEE Journal of Quantum Electronics*, vol. 43, no. 11, pp. 1006–1017, 2007.

[178] C. Gärtner, C. Pflumm, C. Karnutsch, V. Haug, and U. Lemmer, "Numerical study of annihilation processes in electrically pumped organic semiconductor laser diodes," in *Proceedings of SPIE*, ser. Organic Light-Emitting Materials and Devices X, vol. 6333, no. 63331J, San Diego, CA, USA, 2006.

[179] R. Schmechel, M. Ahles, and H. von Seggern, "A pentacene ambipolar transistor: Experiment and theory," *Journal of Applied Physics*, vol. 98, pp. 4511–+, Oct. 2005.

[180] M. Muccini, "A bright future for organic field-effect transistors," *Nature Materials*, vol. 5, pp. 605–613, Aug. 2006.

[181] A. Hepp, H. Heil, W. Weise, M. Ahles, R. Schmechel, and H. von Seggern, "Light-emitting field-effect transistor based on a tetracene thin film," *Phys. Rev. Lett.*, vol. 91, no. 15, p. 157406, Oct 2003.

[182] E. J. Meijer, D. M. de Leeuw, S. Setayesh, E. van Veenendaal, B. H. Huisman, P. W. M. Blom, J. C. Hummelen, U. Scherf, and T. M. Klapwijk, "Solution-processed ambipolar organic field-effect transistors and inverters," vol. 2, no. 10, pp. 678–682, 2003, tY - JOUR 10.1038/nmat978.

[183] R. Schmechel, A. Hepp, H. Heil, M. Ahles, W. Weise, and H. von Seggern, "Light-emitting field-effect transistor: simple model and underlying functional mechanisms," in *Organic Field Effect Transistors II. Edited by Dimitrakopoulos, Christos D. Proceedings of the SPIE, Volume 5217, pp. 101-111 (2003).*, ser. Presented at the Society of Photo-Optical Instrumentation Engineers (SPIE) Conference, C. D. Dimitrakopoulos, Ed., vol. 5217, Nov. 2003, pp. 101–111.

[184] T. Takenobu, S. Z. Bisri, T. Takahashi, M. Yahiro, C. Adachi, and Y. Iwasa, "High current density in light-emitting transistors of organic single crystals," *Physical Review Letters*, vol. 100, no. 6, p. 066601, 2008.

[185] J. Zaumseil, C. Donley, J.-S. Kim, R. H. Friend, and H. Sirringhaus, "Spatial control of the recombination zone in ambipolar light-emitting polymer transistors," *Proceedings of SPIE*, vol. 6192, pp. 1–9, 2006.

[186] J. Zaumseil and H. Sirringhaus, "Electron and ambipolar transport in organic field-effect transistors," *Chemical Reviews*, vol. 107, pp. 1296– 1323, 2007.

[187] J. Zaumseil, R. H. Friend, and H. Sirringhaus, "Spatial control of the recombination zone in an ambipolar light-emitting organic transistor," *Nature Materials*, vol. 5, no. 1, pp. 69–74, 2006.

[188] S. Schols, S. Verlaak, and P. Heremans, "A novel organic light-emitting device for use in electrically pumped lasers," in *Proceedings of SPIE*, vol. 6333, San Diego, CA, USA, 2006, p. 63330U.

[189] S. De Vusser, S. Schols, S. Steudel, S. Verlaak, J. Genoe, W. Oosterbaan, D. Vanderzande, and P. Heremans, "Light-emitting organic field-effect transistor using an organic heterostructure within the transistor channel," *Applied Physics Letters*, vol. 89, p. 223504, 2006.

[190] N. Giebink and S. Forrest, "Accumulation of electric-field-stabilized geminate polaron pairs in an organic semiconductor to attain high excitation density under low intensity pumping," *Applied Physics Letters*, vol. 89, p. 193502, 2006.

[191] ——, "Limits to accumulation of electric-field-stabilized geminate polaron pairs in an organic semiconductor thin film," *Physical Review B*, vol. 76, p. 075318, 2007.

[192] T. Matsushima and C. Adachi, "Extremely low voltage organic light-emitting diodes with p-doped alpha-sexithiophene hole transport and n-doped phenyldipyrenylphosphine oxide electron transport layers," *Applied Physics Letters*, vol. 89, no. 25, p. 253506, 2006.

[193] O. Heavens, *Optical properties of thin solid*. Dover: Dover Publications, 1991.

[194] H. Fujikawa, M. Ishii, S. Tokito, and Y. Taga, "Organic light-emitting diodes using triphenylamine based hole transporting materials," *Mat. Res. Soc. Symp. Proc.*, vol. 621, pp. Q3.4.1–Q3.4.11, 2000.

[195] S. Naka, H. Okada, H. Onnagawa, and T. Tsutsui, "High electron mobility in bathophenanthroline," *Applied Physics Letters*, vol. 76, no. 2, pp. 197–199, 2000.

[196] Y. Shirota, K. Okumoto, H. Ohishi, M. Tanaka, M. Nakao, K. Wayaku, S. Nomura, and H. Kageyama, "Charge transport in amorphous molecular materials," *Proceedings of SPIE*, vol. 5937, no. 593717, 2005.

[197] A. Campbell, D. Bradley, and H. Antoniadis, "Dispersive electron transport in an electroluminescent polyfluorene copolymer measured by the current integration time-of-flight method," *Applied Physics Letters*, vol. 79, no. 14, pp. 2133–2135, 2001.

[198] S. Lee, T. Yasuda, M. Yang, K. Fujita, and T. Tsutsui, "Charge carrier mobility in vacuum-sublimed dye films for light-emitting diodes studied by the time-of-flight technique," *Molecular Crystals and Liquid Crystals*, vol. 405, pp. 67–73, 2003.

[199] H. Yamamoto, T. Oyamada, H. Sasabe, and C. Adachi, "Amplified spontaneous emission under optical pumping from an organic semiconductor laser structure equipped with transparent carrier injection electrodes," *Applied Physics Letters*, vol. 84, no. 8, pp. 1401–1403, 2004.

[200] V. Kozlov, G. Parthasarathy, P. Burrows, V. Khalfin, J. Wang, S. Chou, and S. Forrest, "Structures for organic diode lasers and optical properties of organic semiconductors under intense optical and electrical excitations," *IEEE Journal of Quantum Electronics*, vol. 36, no. 1, pp. 18–26, 2000.

[201] R. Xia, G. Heliotis, Y. Hou, and D. Bradley, "Fluorene-based conjugated polymer optical gain media," *Organic Electronics*, vol. 4, pp. 165–177, 2003.

[202] M. Baldo and S. Forrest, "Transient analysis of organic electrophosphorescence: I. transient analysis of triplet energy transfer," *Physical Review B*, vol. 62, no. 16, pp. 10 958–10 966, 2000.

[203] M. Baldo, "The electronic and optical properties of amorphous organic semiconductors," Dissertation, Princeton University, 2001.

[204] M. Baldo, R. Holmes, and S. Forrest, "Prospects for electrically pumped organic lasers," *Physical Review B*, vol. 66, no. 035321, 2002.

[205] M. Stevens, C. Silva, D. Russell, and R. Friend, "Exciton dissociation mechanisms in the polymeric semiconductors poly(9,9-dioctylfluorene) and poly(9,9-dioctylfluorene-co-benzothiadiazole)," *Physical Review B*, vol. 63, no. 165213, 2001.

[206] J. Yu, R. Lammi, A. J. Gesquiere, and P. F. Barbara, "Singlet-triplet and triplet-triplet interactions in conjugated polymer single molecules," *J. Phys. Chem. B*, vol. 109, pp. 10 025–10 034, 2005.

[207] E. List, U. Scherf, K. Müllen, W. Graupner, C. Kim, and J. Shinar, "Direct evidence for singlet-triplet exciton annihilation in p-conjugated polymers," *Physical Review B*, vol. 66, p. 235203, 2002.

[208] A. Haugeneder, M. Neges, C. Kallinger, W. Spirkl, U. Lemmer, J. Feldmann, M.-C. Amann, and U. Scherf, "Nonlinear emission and recombination in conjugated polymer waveguides," *Journal of Applied Physics*, vol. 85, no. 2, pp. 1124–1130, 1999.

[209] G. D. Hale, S. J. Oldenburg, and N. J. Halas, "Observation of triplet exciton dynamics in conjugated polymer films using two-photon photoelectron spectroscopy," *Physical Review B*, vol. 55, no. 16069, 1997.

[210] S. V. Frolov, M. Liess, P. A. Lane, W. Gellermann, and Z. V. Vardeny, *Phys. Rev. Lett.*, vol. 78, no. 22, 1997.

[211] M. Baldo and S. Forrest, "Interface-limited injection in amorphous organic semiconductors," *Physical Review B*, vol. 64, no. 085201, 2001.

[212] M. Baldo, C. Adachi, and S. Forrest, "Transient analysis of organic electrophosphorescence. ii. transient analysis of triplet-triplet annihilation," *Physical Review B*, vol. 62, no. 16, pp. 10967–10977, 2000.

[213] J. Frenkel, "On pre-breakdown phenomena in insulators and electronic semiconductors," *Phys. Rev.*, vol. 54, no. 8, pp. 647–648, 1938.

[214] Y. A. Cherkasov, "Exciton effects in disordered semiconductors," *Soviet Physics Journal*, vol. 19, pp. 203–214, Feb. 1976.

[215] J. L. Hartke, "The Three-Dimensional Poole-Frenkel Effect," *Journal of Applied Physics*, vol. 39, pp. 4871–4873, Sep. 1968.

[216] C. Gärtner, C. Karnutsch, C. Pflumm, and U. Lemmer, "The influence of annihilation processes on the threshold current density of organic laser diodes," *Journal of Applied Physics*, vol. 101, no. 023107, 2007.

[217] L. V. Asryan and S. Luryi, "Effect of internal optical loss on threshold characteristics of semiconductor lasers with a quantum-confined active region," *IEEE Journal of Quantum Electronics*, vol. 40, pp. 833–843, 2004.

[218] C. Byeon, M. McKerns, W. Sun, and G. Gray, "Excited state lifetime and intersystem crossing rate of asymmetric pentaazadentate porphyrin-like metal complexes," *Applied Physics Letters*, vol. 84, no. 25, pp. 5174–5176, 2004.

[219] R. Higgins, A. Monkman, H.-G. Nothofer, and U. Scherf, "Energy transfer to porphyrin derivative dopants in polymer light-emitting diodes," *Journal of Applied Physics*, vol. 91, no. 1, pp. 99–105, 2002.

[220] S. Naka, H. Okada, H. Onnagawa, Y. Yamaguchi, and T. Tsutsui, "Carrier transport properties of organic materials for el device operation," *Synthetic Metals*, vol. 111-112, pp. 331–333, 2003.

[221] J. Shinar, *Organic Light-Emitting Devices - A Survey*. New York: Springer Verlag, 2004.

[222] I. Campbell, D. Smith, C. Neef, and J. Ferraris, "Consistent time-of-flight mobility measurements and polymer light-emitting diode current-voltage characteristics," *Applied Physics Letters*, vol. 74, no. 19, pp. 2809–2811, 1999.

[223] L. Bozano, S. Carter, J. Scott, G. Malliaras, and P. Brock, "Temperature- and field-dependent electron and hole mobilities in polymer light-emitting diodes," *Applied Physics Letters*, vol. 74, no. 8, pp. 1132–1134, 1999.

[224] M. Baldo, D. O'Brien, Y. You, A. Shoustikov, S. Sibley, M. Thompson, and S. Forrest, "Highly efficient phosphorescent emission from organic electroluminescent devices," *Nature*, vol. 395, pp. 151–154, 1998.

[225] D. E. Klimek, "The effect of triplet quenchers on vapor-phase dye-laser performance," *Applied Physics B*, vol. 34, pp. 83–86, 1984.

[226] N. Tessler, D. Pinner, V. Cleave, D. Thomas, G. Yahioglu, P. L. Barny, and R. Friend, "Pulsed excitation of low-mobility light-emitting diodes: Implication for organic lasers," *Applied Physics Letters*, vol. 74, no. 19, pp. 2764–2766, 1999.

[227] C. Pflumm, C. Gärtner, C. Karnutsch, and U. Lemmer, "Influence of electronic properties on the threshold behaviour of organic laser diode structures," in *Proceedings of SPIE*, ser. Organic Light-Emitting Materials and Devices X, vol. 6333, no. 63330W, San Diego, CA, USA, 2006.

Index

absorption, 12
absorption parameter, 70
accumultion channel, 44
activation scheme, 12
aluminum doped zinc oxide, 59
ambipolar OFET, 43
ambipolar transport, 43
annihilation process, 60
annihilation rate constant, 28
attenuation coefficient, 19, 40, 53
Auger process, 29
AZO, 59

Bässler model, 16
backflow, 20
bandgap, 8, 12
bandgap energy, 12
bimolecular annihilation process, 24, 26,
    27, 56
binding energy, 12
bipolaron, 32
Born-Oppenheimer approximation, 13
breakdown field, 35

carbon, 6
carrier back-flow, 22
cavity, 36
charge carrier, 6, 15
charge injection, 19
charge pair, 12
charge transfer, 24
charge transport, 16
conductance band, 21
conjugated dendrimers, 8, 10
conjugated unit, 10
conjugation, 8

continuity equation, 55
Coulomb binding energy, 12

device geometry, 49
Dexter transfer, 24
dielectric material constant, 12
dipole-dipole transfer, 23
distributed feedback laser, 37
double bond, 6
double-heterostructure, 43
drift-diffusion approximation, 55
drift-diffusion model, 55

effective cross section, 33
effective refractive index, 37
effective Richardson constant, 20
efficiency roll-off, 29
electric field, 34
electrical breakdown field, 35
electrode, 19
electrode quenching, 43
electroluminescence, 43
electron, 15
electron-hole pair, 12
emission, 12
emission layer, 40
ethane, 6, 7
ethene, 6, 7
ethyne, 6, 7
excited state, 37
exciton, 12
exciton accumulation, 44
exciton binding energy, 34
exciton dissociation, 34
exciton formation, 18
excitonic state, 12

Förster formula, 23
Förster radius, 24
Förster transfer, 23
Fabry-Perot cavity, 37
Fermi's golden rule, 23
field emission, 20
field quenching, 34, 56, 63
field-induced exciton dissociation, 26, 34, 56
field-stabilized geminate charge pairs, 44
fluorescence, 13, 18
fluorescent lifetime, 24
fluorescent resonant energy transfer, 23
four-level laser, 37
four-level system, 37
Fowler-Nordheim tunneling, 20
Franck-Condon factors, 13
FRET, 23

gain, 18
gain medium, 36
gate electrode, 44
generation, 55
grating, 39
ground state, 6, 12, 37

Helmholtz equation, 52
Helmholz equation, 52
highest occupied molecular orbital, 12
hole, 15
HOMO, 12
hopping separation theory, 35
hopping transport, 15, 20
hybrid orbitals, 6

ILC, 19
image charge potential, 19
indium tin oxide, 59
inert gas, 39
injection, 19
injection barrier, 19
injection limited current, 19
intercombination, 30
interconversion, 18

interface matrix, 54
intersystem crossing, 14, 31
intrachain mobility, 15
ISC, 31
ITO, 59

Langevin recombination, 18, 22, 56
Langevin theory, 18
LASER, 36
laser, 36
laser resonator, 36
laser transition, 37
lattice distortion, 15
light amplification, 36
linker, 10
low-threshold operation, 36
lowest unoccupied molecular orbital, 12
LUMO, 12

microdisk resonator, 37
microring resonator, 37
microspherical resonator, 37
mobility, 26
modal intensity profile, 52
modelling, 49

nonradiative decay, 12

Onsager theory, 35
optical buffer layer, 41
optical feedback, 36, 37
organic double-heterostructure, 40, 49
organic field-effect transistor, 41
organic laser, 37
organic semiconductor laser, 36, 44

p-orbital, 6
particle flux, 56
phase, 37
phonon assisted tunneling, 15
phosphorescence, 14, 30
photoinduced absorption, 29
photoinduced absorption spectroscopy, 32
photoluminescence, 29

photoluminescence efficiency, 34
photon, 37
photon density, 54
π-conjugated polymers, 8
π-state, 8
π^*-state, 8
PL-detected magnetic resonance, 29
polarization, 37
polaron, 15
polaron absorption, 15, 32
polyethen, 6
polymer chain, 8, 15
polypropene, 6
polystyrene, 6
polyvinyl chloride, 6
Poole-Frenkel approach, 16
Poole-Frenkel theory, 35
population inversion, 36
propagation matrix, 54
pumping, 37

radiative decay, 12
radical anion, 15
radical cation, 15
random diffusion, 18
rate coefficient, 31, 60
rate equation, 54
recombination, 18, 55
refractive index, 53
relative absorption cross section, 70
relaxation, 12, 29
relaxation transition, 37
reverse intersystem crossing, 31
RISC, 31
ruby crystal, 36

s-orbital, 6
saturated bonds, 6
SCLC, 19
Scott-Malliaras modell, 21
σ-bonds, 6
single bond, 6
singlet absorption, 33
singlet annihilation rate term, 62

singlet exciton, 18, 62
singlet-electron annihilation, 29
singlet-polaron annihilation, 29
singlet-singlet annihilation, 27
singlet-triplet annihilation, 30
slab waveguide, 37
small-molecule materials, 8
solution processing, 8
source term, 55
sp^2-hybridisation, 6
sp^3-hybridisation, 6
SPA, 29
space-charge limited current, 19
spin multiplicity, 18
spin statistics, 18
SSA, 27
STA, 30
stimulated emission, 36, 37, 54, 56
stokes shift, 13
surface group, 10

TCO, 19, 59
TE mode, 52
thermal activation energy, 12
thermionic emission, 19
thermionic injection, 22
three-level systems, 36
TM mode, 52
total reflection, 51
TPA, 30
transfer-matrix method, 52
transient photoluminescence, 28
transparent conductive oxide, 19, 41, 59
transport layer, 40
triangular potential barrier, 20
triple bond, 6
triplet exciton, 18, 33, 62
triplet-polaron annihilation, 30
triplet-triplet annihilation, 30
TTA, 30
tunable laser, 37

vibronic states, 12

vibronic subband, 12
vibronic substate, 12

waveguide, 19, 37, 51
waveguide cladding, 51
waveguide core, 51
waveguide loss, 54
wavelength, 37

Nomenclature

α	attenuation coefficient
ϵ_0	vacuum dielectric constant
ϵ_r	dielectric material constant
κ_{ISC}	intersystem crossing annihilation rate constant
κ_{SPA}	singlet-polaron annihilation rate constant
κ_{SSA}	singlet-singlet annihilation rate constant
κ_{STA}	singlet-triplet annihilation rate constant
κ_{TPA}	triplet-polaron annihilation rate constant
κ_{TTA}	triplet-triplet annihilation rate constant
μ	charge carrier mobility
μ_0	zero-field charge carrier mobility
Φ_{A}	work function
Φ_{HOMO}	HOMO energy level
Φ_{LUMO}	LUMO energy level
$\tau_{\mathrm{S_1}}$	singlet exciton lifetime
$\tau_{\mathrm{T_1}}$	triplet exciton lifetime
$\widetilde{n}$	complex refractive index
ξ	singlet exciton generation efficiency
A^*	effective Richardson constant
e	elementary charge
E_{F}	Fermi energy
F	electric field

$I(t)$ photoluminescence intensity

j_{thr} threshold current density

k_{B} Boltzmann constant

k_{Langevin} Langevin rate

n refractive index

N_0 density of chargeable sites

n_{e} electron density

n_{h} hole density

n_{S_1} singlet exciton density

n_{T_1} triplet exciton density

r_{c} Coulomb radius

T temperature

HOMO   highest occupied molecular orbital

LUMO   lowest unoccupied molecular orbital